Mathematics—
The Alphabet of Science

Also by Margaret F. Willerding:

ELEMENTARY MATHEMATICS:
Its Structure and Concepts

John Wiley & Sons, Inc., 1966

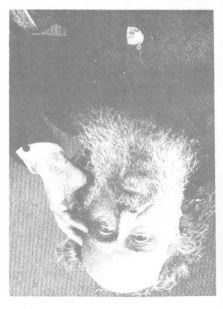

Mathematics
the alphabet of science

Margaret F. Willerding

Professor of Mathematics
San Diego State College

Ruth A. Hayward

Research Engineer
General Dynamics
Convair Division

John Wiley and Sons, Inc.   New York   London   Sydney

PICTURE CREDITS

Bettmann Archive, Chapters
1–7,10; Wide World Photos,
Chapter 8; Scripta Mathe-
matica, Chapter 9; I.B.M.,
Chapter 11

Library of Congress Catalog
Card Number: 68–14097
GB 471 94660X

Printed in the United States
of America

To our mothers,
Mildred Willerding Emme
and Myrtle Byrd McKinney

Preface

     This book is intended for liberal-arts students and other students who wish to know what mathematics is about, but who have no desire to be mathematicians. The presentation is designed so that even persons who have had little or no high-school mathematics, or whose mathematics study is so far in the past that it is completely forgotten, will have no trouble following the explanations.

     The topics selected for discussion are simple yet profound. Many have applications in fields other than mathematics. Explanations have been made as detailed as is reasonably possible. The manner in which the subjects are treated will cause no trouble even to the most mathematically unsophisticated student. For a complete understanding of the topics presented, the reader should verify the calculations and ponder the arguments as he encounters them. But even if the reader reads this book as he reads a novel, he will find enough to acquaint him with some exciting topics of mathematics, both ancient and modern.

     Many of the chapters are completely independent of the others. If there is a particular need for a topic that has been presented in a previous chapter, the sections needed are referred to in a footnote.

     Chapter 1 presents the basic notions of mathematical logic and gives the reader a respect for correct reasoning and a clear understanding of the type of deductive reasoning used in mathematics. Chapters 2, 3, and 4 form a block and should be studied together. All three chapters present topics from elementary number theory. These chapters demand only the mathematics studied in elementary school. Chapter 5 discusses the Pythagorean theorem and related topics. Chapters 6 and 8 are abstract and may be postponed until after the other topics have been studied. Both Chapters 6 and 8 study mathematical systems and stress the postulational method. Chapter 6 gives the reader an opportunity to study a geometry with only a finite number of points and lines, rather than the infinite geometry studied in high school. Chapter 7 is a brief introduction to the fundamental ideas of probability and statistics. Chapter 9 introduces matrix notation and matrix algebra. Chapters 10 and 11 give a brief introduction to computer arithmetic and computer programming. These chapters have been included because of the influence of the electronic computer on today's society—both technical and nontechnical. We do not claim that Chapter 11 will make a programmer of the reader. The intent of this chapter is to acquaint the reader with one phase of the computer.

Our aim in selecting the topics in the book was to choose those topics that are interesting and those that have important applications in the modern world. We chose subjects that we felt were not "old hat" to the students. All topics are developed with the idea that the reader has had no previous knowledge of the subject.

The book is designed for a one-semester course. The material has been class-tested over a period of three years. It is well established that the material is presented in understandable form and is appealing to the student.

We gratefully express our appreciation to William G. Chinn and Carol H. Kipps for critically reading the manuscript and for many helpful suggestions.

Margaret F. Willerding

Ruth A. Hayward

San Diego, California
February 1968

Contents

x   Contents

Mathematics—
The Alphabet of Science

ARISTOTLE (384–322 B.C.), one of the greatest thinkers and scientific investigators of all time, is called the father of traditional (Aristotelian) logic based on the syllogism. He was the first philosopher to divide the branches of philosophy into ethics, logic, metaphysics, physics, politics, and the philosophy of art. Aristotle was not a professional mathematician. He was concerned with mathematics primarily as an illustration of correct reasoning.

Logic / 1

1. HISTORY

The importance of studying the procedures of correct reasoning—that is, **logic**—is obvious to all educated persons. The ancient Greeks developed logic as an area of study, and one of the leaders in this field, Aristotle, categorized arguments according to definite forms. These lines of reasoning are still identified as those of Aristotelian logic.

The growth of mathematics in the nineteenth century caused mathematicians to examine critically the reasoning procedures which had been used in developing mathematics. They discovered that many arguments used by earlier mathematicians were either incomplete or incorrect. Efforts to correct these mistakes and to avoid similar ones renewed interest in logic and resulted in a great extension and development of the study of logic. This new development is called **mathematical** or **symbolic logic** to distinguish it from the old Aristotelian logic.

Traditionally, logic is divided into two parts, **inductive logic** and **deductive logic**. A conclusion reached by repeated observations, as in a series of experiments, illustrates induction; a conclusion based on so-called "if-then" reasoning, as in mathematical proofs, represents deduction. Both Aristotelian logic and symbolic logic are parts of deductive logic. Since the deductive method is an important part of modern mathematical thinking, we shall discuss only deductive logic and refer to it simply as **logic**.

One of the principal concerns of logic is the analysis of the process of establishing the truth of general statements. In mathematics this process is called **proof**. Our objective is to introduce the most fundamental ideas of logic and an understanding of the nature of mathematical proof.

2. DIAGRAMS IN DEDUCTIVE REASONING

The use of diagrams often is helpful in deciding whether our reasoning in an argument is correct. Suppose the following facts are assumed:

All factory workers are union members.
Kramer works in a factory.

What conclusion can be drawn? Figure 1.1 shows that the set of all factory workers (F) are included within the set of union members (U) as our first assumption states. The rectangle in Figure 1.1 represents the set of all people (P); the larger circle represents the set of all union members (U); and the smaller circle represents the set of all factory workers (F). If a person is a factory worker he must be represented by a point in the smaller circle. Moreover, any point outside the smaller circle cannot represent a factory worker. Since the smaller circle falls completely within the larger one, any point in the smaller circle also must be in the larger one. Thus we can conclude that Kramer belongs to a union. We write this **argument** as follows:

All factory workers are union members.
Kramer works in a factory.

∴ Kramer belongs to a union.

The symbol ∴ indicates the conclusion and is read "therefore." Notice that a line separates the **assumptions** from the **conclusion**.

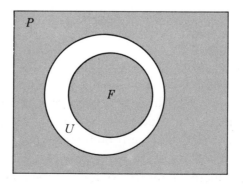

figure 1.1

If the assumptions in the foregoing argument are altered slightly, no conclusions can be drawn. Suppose we assume the following:

> All factory workers are union members.
> Kramer does not work in a factory.

Using Figure 1.1, we see that no valid conclusion can be drawn. A person who is not a factory worker may or may not be a union member.

The following examples illustrate the use of diagrams to check the validity of an argument.

Example 1.

> All mathematicians are intelligent.
> Some women are not intelligent.
>
> ─────────────
>
> ∴ Some women are not mathematicians.

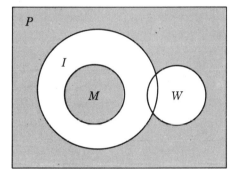

Observe that we have taken care of each assumption in the diagram. The first assumption states that the *M* circle must be included within the *I* circle. The second assumption states only that there are points in the *W* circle outside the *I* circle. There are two possible placements for the *W* circle as shown in the following diagrams.

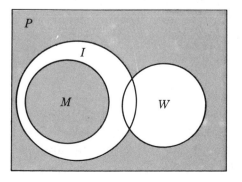
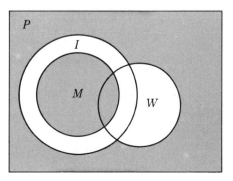

Notice that in each of these cases, the conclusion may be accepted.

Diagrams are helpful in deciding the validity of a conclusion, but they must be used with caution. For example, a nonvalid conclusion from the foregoing assumptions is: no mathematicians are women. Not enough information is given to reach this conclusion.

Example 2.

> No ooks are acks.
> All kooks are ooks.
> All pocks are acks.
> ───────────
> ∴ No kooks are pocks.

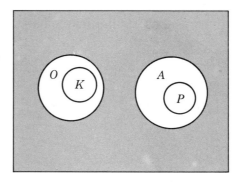

The foregoing diagram shows the validity of the conclusion. Other conclusions that can be drawn from the assumptions are:

> No kooks are acks.
> All pocks are not kooks.

This argument shows that one may reason correctly even though the assumptions have no meaning. An argument in which a conclusion is reached correctly is said to be **valid**. The conclusion may be true or it may be false. The validity of the argument has nothing to do with the truth of the assumptions or the conclusion.

3. UNDEFINED TERMS

The first requirement for an understanding of any subject, be it golf or mathematics, is to know the meaning of the terms that are used. As children we learned the meaning of a word, such as "cat," when someone pointed to an animal and called it "cat." Later we acquired the habit of looking up the definitions of unfamiliar words in the dictionary. A little experience in using a dictionary convinces us that some words must be **undefined**. Unless we understand some words, a dictionary will lead us in a circle. For example, suppose we look up the adjective "stubborn." We may find that, as we look up each definition given, we are ultimately led back to "stubborn."

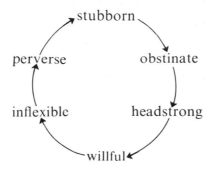

To avoid circular definitions we must accept a small number of words that will be undefined. The choice of these words depends on the subject. With these basic undefined words and nontechnical words from the English language, we now define other words in terms of our undefined words and nontechnical English words. For example, if we accept "point," "line," and "betweenness" as undefined words, we may now define the line segment PQ, where P and Q are two points on a line, as the set consisting of the points P and Q and all the points on the line between P and Q.

4. STATEMENTS

With our undefined words, nontechnical words, and the words we defined, we form sentences called **statements** or **propositions**. These statements must have a truth value. That is, they must be either true or false, but not both. The following sentences are statements.

> A cow has four legs.
> Elephants can fly.
> Hawaii is the fiftieth state of the U.S.A.
> Water is wet.
> July has thirty days.
> Two is an odd number.

The following sentences are not statements by our definitions because as they stand they have no truth value. That is, we cannot tell whether they are true or false.

> That number is divisible by 6.
> He is an honor student in mathematics.
> She is fifteen pounds overweight.

In a deductive system such as mathematics we accept a set of statements, called **axioms** or **postulates**, as true. Once these initial propositions are stated, we combine them into more complicated statements. We attempt to prove these complicated statements true by a process called **logical reasoning**. Those propositions that turn out to be true are called **theorems**.

EXERCISE 1

1. Which of the following are statements? Give the truth value of each statement.
   (a) January is the first month of the year.
   (b) She is sixteen years old.
   (c) Texas is the largest state in the United States.
   (d) All cats have lavender eyes.
   (e) Ice is cold.
   (f) Buttermilk is a dairy product.
   (g) He is a juvenile delinquent.
   (h) The sum of 6 and 9 is 12.
   (i) The diameter of the earth is 1,000 miles.
   (k) That number is a prime number.
2. Using diagrams, tell what conclusion can be drawn from the following sets of assumptions.
   (a) No even number is divisible by 3.
       Six is an even number.
   (b) If a boy owns a car, he does not get A grades.
       John owns a car.
   (c) All rational numbers are real numbers.
       All real numbers can be graphed on the real number line.
3. Using diagrams, tell what conclusion can be drawn from each set of assumptions.
   (a) Some girls are talkative.
       Talkative people are not popular.

    (b) All irrational numbers are real numbers.

       $\sqrt{2}$ is an irrational number.

    (c) All girls are beautiful.

       All career women are girls.

4. Using diagrams tell what conclusion can be drawn from each set of assumptions.

    (a) All Californians drink wine.

       Doris lives in California.

    (b) All peeps are creeps.

       No parfaits are creeps.

    (c) All rings are groups.

       All modular systems are rings.

5. Using diagrams tell what conclusion can be drawn from each set of assumptions.

    (a) All girls have two ears.

       Meg is a girl.

    (b) All handsome men are college students.

       No freshmen are college students.

    (c) No excellent talkers are bores.

       Some excellent talkers are college professors.

6. Test the validity of the following argument by using diagrams.

  All mathematicians are bores.

  All bores are stupid.

  _____

  ∴ All mathematicians are stupid

7. Test the validity of the following arguments by using diagrams.

  All shy creatures are finks.

  Some shy creatures are rats.

  Some students are shy creatures.

  _____

  ∴ Some students are rat finks.

8. Draw a diagram for each of the following arguments. Use the diagram to tell whether the conclusion of each argument follows from the assumptions.

    (a) All church-goers are good people.

       Mrs. Lehman goes to church.

      _____

    ∴ Mrs. Lehman is a good person.

    (b) All San Franciscans are residents of California.

       All Hippies are residents of California.

      _____

    ∴ All Hippies are San Franciscans.

9. Draw a diagram for each of the following arguments. Use the diagram to tell whether the conclusion of each argument follows from the assumptions.
   (a) All women are intelligent animals.
   My cat, Madame Nu, is an intelligent animal.
   ────────────
   ∴ My cat is a woman.
   (b) All beer drinkers have red noses.
   Tom has a red nose.
   ────────────
   ∴ Tom is a beer drinker.
10. Draw a diagram for each of the following arguments. Use the diagram to tell whether the conclusion of each argument follows from the assumptions.
   (a) All senators are over twenty-one years of age.
   Fred is over twenty-one years of age.
   ────────────
   ∴ Fred is a senator.
   (b) All even numbers are integers.
   All odd numbers are integers.
   ────────────
   ∴ All odd numbers are even numbers.

5. COMPOUND STATEMENTS

Logical reasoning is the process of combining given statements into other statements, and then doing this over and over and over again. There are many ways to combine statements, all of which are derived from four fundamental operations: (1) conjunction, (2) disjunction, (3) implication, and (4) negation.

In a **conjunction** we combine two given statements by placing an "and" between them. Thus the conjunction of the two statements

(a) Ross is a physics major.
(b) Wilson is a mathematics major.

is "Ross is a physics major and Wilson is a mathematics major."

A conjunction is considered true if both of the statements, called **components**, used to form it are true. Thus, if either component is false, the conjunction is defined as false. Of course, if both components are false, the conjunction likewise is false.

It is customary to denote statements by lower-case letters of the alphabet such as p, q, r, and s. If p and q represent the components used to form a conjunction, we write

$$p \wedge q$$

and read this "p and q." Thus we can summarize the truth values of p and q:

> if p is true and q is true, then $p \wedge q$ is true
> if p is false and q is true, then $p \wedge q$ is false
> if p is true and q is false, then $p \wedge q$ is false
> if p is false and q is false, then $p \wedge q$ is false

This information can be given more compactly in the form of a **truth table**. Table 1.1 is a truth table giving the truth values of the conjunction. In the table, T means that the corresponding statement is true; F, that it is false. This table is merely an expansion of our description of conjunction. It says that if both p and q are true, then $p \wedge q$ is true; if one or the other or both p and q are false, then $p \wedge q$ is false.

TABLE 1.1

p	q	$p \wedge q$
T	T	T
T	F	F
F	T	F
F	F	F

To form a truth table all possible combinations of true and false are entered in the p and q columns, and the truth or falsity corresponding to each combination is entered in the third column.

In a **disjunction** we combine two statements by placing "or" between them. The disjunction of the preceding statements (a) and (b) is "Ross is a physics major or Wilson is a mathematics major." The disjunction is defined to be true if either one or the other or both of its components are true and false otherwise. Notice that the connective "or" when used in a disjunction is used in the **inclusive** sense; that is, "or" means one component or the other or both. In everyday language we often use "or" in the **exclusive** sense; that is, "or" means one component or the other but not both. For example, when a student says "I shall get an A or a B in this course" he means that he will get an A or a B but not both grades in the course. Hereafter, the word "or" will always be used in this book in the inclusive sense.

If p and q are the components used to form the disjunction, we write

$$p \lor q$$

and read this "p or q." Table 1.2 shows the truth values of the disjunction $p \lor q$.

TABLE 1.2

p	q	$p \lor q$
T	T	T
T	F	T
F	T	T
F	F	F

6. NEGATION

If p is any statement, the statement "not p" or "p is false" is called the **negation** of p. We denote this by the symbol $\sim p$. For example, if p denotes the statement "Fred likes mathematics," then $\sim p$ denotes the statement "It is false that Fred likes mathematics" or "Fred does not like mathematics."

Clearly, p and $\sim p$ have opposite truth values. This fact is shown by the entries in Table 1.3.

TABLE 1.3

p	$\sim p$
T	F
F	T

7. IMPLICATION

A very important compound statement in mathematics is the **implication**. It takes the form of an "if-then" sentence. The following statements are implications.

> If I study hard, then I shall pass this course.
> If it rains, then the picnic will be canceled.
> If I get paid, then I shall go to the movies tonight.

From any two statements we can form two implications. From "It is raining" and "There is a rainbow," we can form the implications:

> If it is raining, then there is a rainbow.
> If there is a rainbow, then it is raining.

In the implication "If it is raining, then there is a rainbow," "it is raining" is the **antecedent**, and "there is a rainbow" is the **conclusion**. In the implication "If there is a rainbow, then it is raining," "there is a rainbow" is the antecedent and "it is raining" is the conclusion. In ordinary speech it is customary for the antecedent and the conclusion to be related, as:

> If I go skiing, I may break my leg.

It is generally meaningless to combine two apparently unrelated statements, as:

> If $2 \times 2 = 4$, then the sky is blue.

In mathematical logic we shall not hesitate to combine unrelated statements in an implication.

If p and q are symbols denoting two statements, we use the symbol $p \rightarrow q$ to denote the **implication** and read it "if p, then q." The symbol $p \rightarrow q$ may also be read "p implies q."

Table 1.4 defines the truth value of the implication $p \rightarrow q$. The first

TABLE 1.4

p	q	$p \rightarrow q$
T	T	T
T	F	F
F	T	T
F	F	T

line in the table is easily accepted. In fact, this may be precisely our interpretation of the implication "$p \rightarrow q$": if p is true, then q is true. The second line is intuitively easy to accept: from the understanding that "if p is true, *then q is true*," it must be false to have p to be true and q to be false. The third and fourth lines are less familiar and more difficult to accept. We can see the reasoning in the fourth line, however, by turning the statement around thus: if it is true that "if p is true, then q is true," then it must be equally true that "if q is false, then p cannot be true"; that is, "p is false and q is false" must be a true statement.

We shall accept the truth table as the definition of the truth values of the implication $p \rightarrow q$. That is, we define the implication $p \rightarrow q$ as true unless p is true and q is false, in which case it is false.

8. EQUIVALENCE

The last of our connectives, "if and only if," symbolized by $p \leftrightarrow q$ and read: p if and only if q, is very important in mathematics. The truth values associated with $p \leftrightarrow q$, called the **biconditional** or the **equivalence**, are given in Table 1.5. From the table we see that the equivalence $p \leftrightarrow q$ is true if p and q have the same truth values, and false if p and q have opposite truth values.

TABLE 1.5

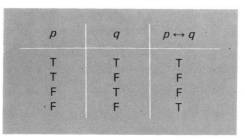

p	q	$p \leftrightarrow q$
T	T	T
T	F	F
F	T	F
F	F	T

EXERCISE 2

1. Form the (1) disjunction and (2) conjunction of the following pairs of statements.
   (a) Hawaii is the newest state in the union.
       Rhode Island is the smallest state in the union.

(b) Today is Tuesday.
Yesterday was Monday.
(c) Two is an even number.
Three is a prime number.
(d) All squares are rectangles.
All rectangles are quadrilaterals.
(e) Chicago is called the Windy City.
New York is called the Empire City.
(f) Addition and subtraction are inverse operations.
Subtraction is more difficult than addition.

2. Use the following symbolic notation:
$$p: \text{the integer } n \text{ is an even number}$$
$$q: \text{the integer } n \text{ is a composite number.}$$
Write each of the following in symbolic notation.
(a) The integer n is an even number and a composite number.
(b) The integer n is an even number or a composite number.
(c) The integer n is not an even number.

3. Given: p: Mexico is south of the U.S.A.
q: Canada is a democracy.
Write the English statements for the following.
(a) $p \rightarrow q$ (c) $\sim p$ (e) $(\sim p) \rightarrow (\sim q)$
(b) $p \leftrightarrow q$ (d) $p \rightarrow (\sim q)$ (f) $(\sim q) \rightarrow p$

4. Give the truth value of each of the following statements.
(a) All birds have wings and dogs have four legs.
(b) All cows have purple eyes and rats have tails.
(c) Two is an even number or 5 is divisible by 3.
(d) A right triangle is a triangle and a square is a triangle.

5. Form the negation of the following statements.
(a) It is foggy today.
(b) The waves are six feet high.
(c) A circle has a radius.
(d) An even number is divisible by 2.
(e) People work for a living.
(f) Yorkshire terriers are intelligent dogs.

6. Give the antecedent and the conclusion of the following implications.
(a) If I work hard, I shall pass this course.
(b) If a triangle is isosceles, then it has two congruent sides.
(c) If a polygon is a quadrilateral, then it has four sides.
(d) If an integer is even, then it has a factor 2.
(e) If the prices are rising, inflation is coming.

7. Let p denote "It is raining," and q denote "The sky is gray."

Translate the following into symbolic notation.
(a) If it is raining, then the sky is gray.
(b) If the sky is gray, then it is raining.
(c) It is raining if and only if the sky is gray.
(d) If the sky is gray, then it is not raining.
(e) It is not raining if the sky is not gray.

8. Which of the following implications are true?
   (a) If $2 + 2 = 4$, then $7 \times 9 = 48$.
   (b) If a triangle is a square, then a rectangle is a pentagon.
   (c) If Lincoln was five feet tall, then John F. Kennedy was president of the United States.
   (d) If cats purr, then dogs bark.
   (e) If Hawaii is not a state, then Alabama is in the north.

9. Which of the following equivalences are true?
   (a) Six is 2×3 if and only if $3 \times 3 = 9$.
   (b) New York is a city if and only if Chicago is a state.
   (c) The President of the United States is less than twenty-one years of age if and only if bears are citizens of the United States.
   (d) A bluejay has red feathers if and only if all dogs read magazines.
   (e) A mouse is a rodent if and only if a dog is a canine.

10. Write the following statements in symbolic form.
    (a) Leo likes Elaine. (Statement p.)
    (b) Elaine likes Leo. (Statement q.)
    (c) Elaine and Leo like each other.
    (d) Elaine likes Leo but Leo dislikes Elaine.
    (e) If Elaine likes Leo then Leo likes Elaine.
    (f) Elaine and Leo dislike each other.

11. Let p be "Property taxes are high" and let q be "Taxes are rising." Give an English statement for each of the following.
    (a) $p \wedge q$ (c) $p \rightarrow q$ (e) $\sim(p \vee q)$
    (b) $p \wedge (\sim q)$ (d) $\sim(p \wedge q)$ (f) $\sim(\sim p \wedge q)$

12. Write the following in symbolic form.
    (a) This quadrilateral is a rectangle. (Statement p.)
    (b) This quadrilateral is a square. (Statement q.)
    (c) If this quadrilateral is a rectangle then it is a square.
    (d) If this quadrilateral is not a square then it is a rectangle.
    (e) This quadrilateral is a square or it is a rectangle.

13. Translate each of the following into English if p is "Nellie is lucky," q is "Joan is intelligent," and r is "Betty is stuffy."

(a) $p \wedge (\sim q)$ (c) $[p \vee (\sim q)] \wedge r$

(b) $(\sim p) \vee (q \wedge r)$ (d) $\sim [(p \vee q) \wedge r]$

14. Write the following in symbolic form. Let p be "Deane is made of money" and let q be "Money is made through hard work."

(a) Deane is not made of money but money is made through hard work.

(b) Either Deane is made of money or money is not made through hard work.

(c) It is not true that Deane is made of money.

(d) If money is made through hard work, then Deane is not made of money.

9. TAUTOLOGY

If a compound statement is true for all possible truth values of its components, it is called a **tautology**. For example, the statement

$$\sim [p \wedge (\sim p)]$$

is a tautology. Table 1.6 demonstrates that this statement is a tautology.

TABLE 1.6

p	$\sim p$	$p \wedge (\sim p)$	$\sim [p \wedge (\sim p)]$
T	F	F	T
F	T	F	T

In constructing the truth table to show that $\sim [p \wedge (\sim p)]$ is a tautology, we use the headings p, $\sim p$, $p \wedge (\sim p)$, and $\sim [p \wedge (\sim p)]$. We write all the possible truth values, T and F, under p and carry each line across to the right inserting the correct truth values. We find $\sim [p \wedge (\sim p)]$ is true for all possible truth values of p, and hence is a tautology. This tautology is called the **Law of Contradiction**.

Another tautology used in logical reasoning is the **Law of the Excluded Middle**. This law states that for every statement p, either p is true or $\sim p$ is true; that is, $p \vee (\sim p)$. Table 1.7 shows that $p \vee (\sim p)$ is a tautology.

TABLE 1.7

p	$\sim p$	$p \vee (\sim p)$
T	F	T
F	T	T

Another important tautology is

$$[(p \to q) \wedge (q \to r)] \to (p \to r).$$

This is called the **Law of Syllogisms**. Table 1.8 verifies that the law of syllogisms is a tautology.

TABLE 1.8

p	q	r	$p \to q$	$q \to r$	$(p \to q) \wedge (q \to r)$	$p \to r$	$[(p \to q) \wedge (q \to r)] \to (p \to r)$
T	T	T	T	T	T	T	T
T	T	F	T	F	F	F	T
T	F	T	F	T	F	T	T
T	F	F	F	T	F	F	T
F	T	T	T	T	T	T	T
F	T	F	T	F	F	T	T
F	F	T	T	T	T	T	T
F	F	F	T	T	T	T	T

To construct Table 1.8 we begin with three columns as follows because three statements are used to form the law of syllogisms. The eight lines show the possible combinations of truth values for p, q, and r.

p	q	r
T	T	T
T	T	F
T	F	T
T	F	F
F	T	T
F	T	F
F	F	T
F	F	F

On the basis of the truth values of p, q, and r we find the truth values of $p \rightarrow q$, $q \rightarrow r$, $(p \rightarrow q) \wedge (q \rightarrow r)$, and $p \rightarrow r$ as shown in Table 1.8. The law of syllogisms is also called the **chain rule**. Successive applications of the rule permit a chain of implications of any desired length. For example, if $p \rightarrow q$ and $q \rightarrow r$ and $r \rightarrow s$ and $s \rightarrow t$, then $p \rightarrow t$.

Using truth tables we can derive a series of tautologies known as the **Rules of Inference**. Some of the more important rules of inference were derived in the foregoing paragraphs.

10. DERIVED IMPLICATIONS

From the implication $p \rightarrow q$ we can form several derived implications, which may or may not be true when the given implication is true. The most important ones are:

Converse	$q \rightarrow p$
Inverse	$(\sim p) \rightarrow (\sim q)$
Contrapositive	$(\sim q) \rightarrow (\sim p)$

Comparing truth tables for $p \rightarrow q$ and $q \rightarrow p$ (Table 1.9), we see that the

TABLE 1.9

p	q	$p \rightarrow q$	$q \rightarrow p$
T	T	T	T
T	F	F	T
F	T	T	F
F	F	T	T

converse may be false when the given implication is true. For example, the implication:

If a quadrilateral is a rectangle, then it is a parallelogram.

is true, but its converse:

If a quadrilateral is a parallelogram, then it is a rectangle.

is false.

Sometimes the converse of a true implication may be true. For example, both the following implication and its converse are true.

Implication: If it is 8 o'clock P.S.T. in San Diego, then it is 10 o'clock C.S.T. in St. Louis.

Converse: If it is 10 o'clock C.S.T. in St. Louis, then it is 8 o'clock P.S.T. in San Diego.

Let us construct a truth table (Table 1.10) to investigate whether the inverse is true when the given implication is true and false when it is false. Since the last two columns of Table 1.10 differ, we can conclude that the inverse of a given implication is not always true when the given implication is true.

TABLE 1.10

p	q	$\sim p$	$\sim q$	$p \rightarrow q$	$(\sim p) \rightarrow (\sim q)$
T	T	F	F	T	T
T	F	F	T	F	T
F	T	T	F	T	F
F	F	T	T	T	T

Now we consider the truth table (Table 1.11) of the contrapositive of a given implication. Since the last two columns of the table are identical,

TABLE 1.11

p	q	$\sim p$	$\sim q$	$p \rightarrow q$	$(\sim q) \rightarrow (\sim p)$
T	T	F	F	T	T
T	F	F	T	F	F
F	T	T	F	T	T
F	F	T	T	T	T

we conclude that an implication and its contrapositive are simultaneously true or false; that is, they are **logically equivalent**. Likewise, the inverse and the converse are logically equivalent. Any two statements that are simultaneously true or false are called logically equivalent or equivalent statements.

EXERCISE 3

1. Construct a truth table for $[\sim(p \lor q)] \lor [(q \lor p)]$.
2. Construct a truth table for $p \rightarrow (q \lor r)$.
3. Construct a truth table for $p \land (\sim p)$.
4. Construct a truth table for $[p \lor (\sim q)] \land r$.
5. Construct a truth table for $(p \rightarrow q) \lor [p \rightarrow (\sim p)]$.
6. Construct a truth table for $(p \rightarrow q) \leftrightarrow [(\sim p) \lor q]$.
7. Using a truth table show that $(p \rightarrow q) \rightarrow \sim[p \land (\sim q)]$ is a tautology.
8. Using a truth table show that $(p \lor q) \leftrightarrow \sim[(\sim p) \land (\sim q)]$ is a tautology.
9. Using a truth table show that $(p \land q) \rightarrow p$ is a tautology.
10. Using a truth table show that $(p \land q) \rightarrow (p \lor q)$ is a tautology.
11. Using a truth table show that $[p \land (\sim p)] \rightarrow q$ is a tautology.
12. Form the (1) converse, (2) inverse, and (3) contrapositive of the following implications.
    (a) If n is an integer, then n is divisible by 1.
    (b) If some college students are communists, then all college students are communists.
    (c) If there is a depression, then prices go down.
    (d) If a triangle is equilateral, then all of its angles have the same measure.
    (e) If a polygon has five sides, then it is a pentagon.
    (f) New Year's Day falls on Monday if Christmas falls on Monday.
    (g) If you drive a car, then you have a driver's license.
    (h) All girls are beautiful if a man is slightly tipsy.
13. Using truth tables show that $(\sim p) \lor (\sim q)$ is logically equivalent to $\sim(p \land q)$.
14. Using truth tables show that $p \land (\sim q)$ and $\sim(p \rightarrow q)$ are logically equivalent.
15. Using truth tables show that $p \rightarrow (\sim q)$ and $(\sim p) \lor (\sim q)$ are logically equivalent.
16. Using truth tables show that $(\sim p) \land (\sim q)$ and $\sim(p \lor q)$ are logically equivalent.
17. Using truth tables show that $p \rightarrow q$ and $(\sim p) \lor q$ are logically equivalent.
18. Is the inverse of the converse of an implication the same as the converse of the inverse of that implication?
19. If p and q are true and r is false, what is the truth value of:
$$[p \lor (\sim q)] \land (\sim r)$$

20. If p is true what is the truth value of the following.
   (a) $p \lor q$ (c) $\sim p$
   (b) $(\sim p) \land q$ (d) $\sim p \rightarrow p$

21. Prove that the conjunction of a statement with itself is logically equivalent to the statement.

22. Prove that the disjunction of a statement with itself is logically equivalent to the statement.

23. Prove that $\sim(\sim p)$ is logically equivalent to p.

11. LAWS OF SUBSTITUTION AND DETACHMENT

Two more laws of logic that are used in deductive reasoning are the law of detachment and the law of substitution.

The **Law of Substitution** states that we may substitute at any point in the deductive process one statement for an equivalent statement. Thus if we know that $p \rightarrow q$ is true, we may at any point substitute the contrapositive, $(\sim q) \rightarrow (\sim p)$, because they are logically equivalent statements.

The **Law of Detachment** states that if an implication $p \rightarrow q$ is true and if the antecedent, p, is true, then the conclusion, q, is true. The proof of the law of detachment follows immediately from Table 1.4. Since we know $p \rightarrow q$ is true we are in lines 1, 3, or 4 of the table. Since p is also true we are in lines 1 or 2 of the table. Line 1 is the only one in which p and $p \rightarrow q$ are true as given; hence q is true.

12. VALID ARGUMENTS

One of the most important tasks of a logician is the checking of the validity of arguments. By an argument we mean the assertion that a certain statement, called the **conclusion**, follows from other statements, called **premises**. An argument is **valid** if and only if the conjunction of the premises implies the conclusion. In other words, if all the premises are true, the conclusion is also true.

It is important to realize that the truth of the conclusion is irrelevant as a test for the validity of the argument. A true conclusion does not necessarily assure a valid argument, nor does a valid argument necessarily assure a true conclusion. The following examples show this. They also illustrate the form in which we write arguments.

Example 1.

> If all triangles are polygons, then the sides of the triangle
> are line segments.
> The sides of the triangle are line segments.

> ∴ All triangles are polygons.

The conclusion is true but the argument is invalid, since the conclusion does not follow from the premises.

Example 2.

> January is hotter than August.
> August is hotter than June.

> ∴ January is hotter than June.

Here the conclusion is false, but the argument is valid since the conclusion follows from the premises. Notice that the first premise is false.

Example 3.

> All owls have four legs.
> All four-legged animals are birds.

> ∴ All owls are birds.

Here the argument is valid and the conclusion is true, but both the premises are false.

Each of the foregoing examples emphasizes the fact that neither the truth value nor the content of the statements in arguments affect the validity of an argument. If the premises are true and the argument is valid, then the conclusion cannot be denied.

Two valid forms of argument are

$$p \rightarrow q \qquad p \rightarrow q$$
$$p \qquad\qquad \sim q$$

$$\therefore q \qquad\qquad \therefore \sim p$$

The argument on the left is true because of the law of detachment. The truth of the argument on the right is shown in the last line of Table 1.12.

TABLE 1.12

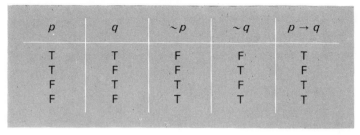

p	q	$\sim p$	$\sim q$	$p \to q$
T	T	F	F	T
T	F	F	T	F
F	T	T	F	T
F	F	T	T	T

The construction of a truth table is one way to check the validity of an argument. Consider the following argument:

$$p \leftrightarrow q$$
$$p$$

$$\therefore q$$

The truth table for the argument is:

p	q	$p \leftrightarrow q$
T	T	T
T	F	F
F	T	F
F	F	T

Both premises are true only in the first row of the table. In this row the conclusion is also true, so the argument is valid.

EXERCISE 4

1. Test the validity of the following arguments.
   (a) No even number is exactly divisible by 7.
   Sixteen is an even number.

   $\therefore$ Sixteen is not exactly divisible by 7.
   (b) All vehicles have four wheels.
   A motor scooter is a vehicle.

∴ A motor scooter has four wheels.

2. Test the validity of the following arguments.
   (a) If today is Wednesday, three days from today will be Saturday.
   Today is not Wednesday.
   _____
   ∴ Three days from today will be Saturday.
   (b) If John passes the final examination, he will pass the course.
   John passes the final examination.
   _____
   ∴ John will pass the course.

3. Test the validity of the following arguments.
   (a) If Sally lives in Los Angeles, then she lives in California.
   Sally lives in California.
   _____
   ∴ Sally lives in Los Angeles.
   (b) If prices are rising, inflation is coming.
   Inflation is coming.
   _____
   ∴ Prices are rising.

4. Test the validity of the following arguments.
   (a) 1964 was a leap year.
   All leap years are election years.
   _____
   ∴ 1964 was an election year.
   (b) All intelligent men are Republicans.
   All Republicans are wealthy.
   _____
   ∴ All intelligent men are wealthy.

5. Test the validity of the following arguments.
   (a) All mathematics books are interesting.
   All comic books are interesting.
   _____
   ∴ All mathematics books are comic books.
   (b) All men are handsome.
   All husbands are men.
   _____
   ∴ All husbands are handsome.

6. Using the law of detachment draw a valid conclusion from the following premises.
   (a) If it is 9 o'clock M.S.T. in Phoenix, it is 8 o'clock P.S.T. in San Francisco. It is 9 o'clock M.S.T. in Phoenix.
   (b) If 1972 is a leap year, it is an election year. 1972 is a leap year.

7. Using the law of detachment draw a valid conclusion from the following premises.

   (a) If X is the murderer, then he was at the scene of the crime. X is the murderer.

   (b) If *ABCD* is a square, then it is a rectangle. *ABCD* is a square.

8. Using the law of detachment draw a valid conclusion from the following premises.

   (a) x is less than y if w is greater than z. w is greater than z.

   (b) If all boys are intelligent, then Frank has two heads. All boys are intelligent.

9. Fill in the blanks with statements to make the following arguments valid.

   (a) If n is an integer, it is either even or odd.

   . . . .

   _____

   $\therefore$ n is either even or odd.

   (b) If $x > y$ and $y > z$, then $x > z$.

   $x > y$ and $y > z$.

   _____

   . . . .

10. Fill in the blanks with statements to make the following arguments valid.

    (a) If n is a natural number, n is a whole number.

    . . . .

    _____

    $\therefore$ n is a whole number.

    (b) . . . .

    I eat too much.

    _____

    $\therefore$ I shall get fat.

11. Fill in the blanks with statements to make the following arguments valid.

    (a) . . . .

    I have a 3.5 grade-point average.

    _____

    $\therefore$ I shall graduate with honors.

    (b) If all mumbos are jumbos, then all booboos are kookoos.

    . . . .

    _____

    $\therefore$ All booboos are kookoos.

12. Construct a truth table to show that the following arguments are invalid.

(a) $p \rightarrow q$

$\sim p$

_____

$\therefore \sim q$

(b) $q \rightarrow (\sim p)$

$\sim p$

_____

$\therefore \sim q$

13. DIRECT PROOF

There are many kinds of proofs: proof by performance, proof by induction, photographer's proof, legal proof, printer's proof, and hundred proof. Mathematical proof is none of these. Mathematical proof is proof by deduction. It consists of showing that the statement to be proved is a logical consequence of given premises.

In a **direct proof** a chain of syllogisms is arranged from the given premises to the desired conclusion.

We learn a good deal about the techniques of proof by reading proofs that others have made. Study the following examples of direct proofs.

Example 1.

It is raining.	p
If the ground is dry, then it is not raining.	$q \rightarrow (\sim p)$
If the ground is wet, then the road is slippery.	$(\sim q) \rightarrow r$

_____   _____

$\therefore$ The road is slippery. $\therefore r$

Proof: Since $q \rightarrow (\sim p)$ is true, the contrapositive $p \rightarrow (\sim q)$ is true. We now have a chain of syllogisms from p to r:

$$p \rightarrow (\sim q)$$
$$(\sim q) \rightarrow r$$

We now know from the law of syllogisms that $p \rightarrow r$. But, since $p \rightarrow r$ is true and p is true, r is true from the law of detachment.

One of the most commonly used forms of logical arguments is presented in Example 2.

Example 2.

Either Gay or Suzie is president.	$p \vee q$
Suzie is not president.	$\sim q$
∴ Gay is president.	∴ p

Proof: Since $\sim q$ is true, q is false. Since $p \vee q$ is true and q is false, p must be true.

Example 3.

If I study hard, then I shall graduate. $p \rightarrow q$
If I don't pass my exams, then I shall not graduate. $(\sim r) \rightarrow (\sim q)$

∴ If I don't pass my exams, then I don't study hard.   ∴ $(\sim r) \rightarrow (\sim p)$

Proof: Since $p \rightarrow q$ is true, the contrapositive, $(\sim q) \rightarrow (\sim p)$, is true. We now have a chain of syllogisms:

$$(\sim r) \rightarrow (\sim q)$$
$$(\sim q) \rightarrow (\sim p)$$

Hence, from the law of syllogisms, $(\sim r) \rightarrow (\sim p)$.

Example 4. If this is a good book, it is worth reading. Either the reading is easy, or the book is not worth reading. The reading is not easy. Therefore this book is not worth reading.

Proof: Let p: This is a good book.
          q: The book is worth reading.
          r: The reading is easy.
Then the premises are

$$p \rightarrow q$$
$$r \vee (\sim q)$$
$$\sim r$$

and the conclusion is $\sim q$.
        Since $\sim r$ is true, r is false. Since r is false and $r \vee (\sim q)$ is true, $\sim q$ is true.

14. INDIRECT PROOF

The **indirect** method of proof is often called proof by contradiction. This method of proof relies on the fact that if $(\sim p)$ is false, p is true. Hence to prove p is true, we attempt to show that $(\sim p)$ is false. The best way to do this is to show that $(\sim p)$ is not consistent with the given premises. In other words, we add $(\sim p)$ to the list of premises and show that with this premise added we have a contradiction. When this contradiction is reached, we know that our assumption is not true; that is, $(\sim p)$ is false and p is true.

Study the following indirect proofs.

Example 1.

$$p$$
$$q \to (\sim p)$$
$$(\sim q) \to s$$
$$\overline{}$$
$$\therefore s$$

Proof: Assume $(\sim s)$ is true. Since $(\sim q) \to s$ is true, the contra-positive, $(\sim s) \to q$ is true. We now have a chain of syllogisms:

$$(\sim s) \to q$$
$$q \to (\sim p)$$

Hence $(\sim s) \to (\sim p)$ by the law of syllogisms.

Now $(\sim s) \to (\sim p)$ is true and $(\sim s)$ is true by our assumption. Hence, by the law of detachment, $(\sim p)$ is true. But this leads to a contradiction because by the premises p is true. Hence our assumption is false, and s is true.

Example 2.

$$p \vee q$$
$$\sim q$$
$$\overline{}$$
$$\therefore p$$

Proof: Assume $(\sim p)$ is true. Since, by assumption, $(\sim p)$ is true, p is false. Since p is false and $p \vee q$ is true, q must be true. But this leads to a contradiction since $(\sim q)$ is given true. Hence our assumption is false and p is true.

Example 3.

Fisher is unemployed. He is on welfare. If Fisher is not unemployed, he has money in the bank. If Fisher is on welfare, he does not have money in the bank. Therefore Fisher does not have money in the bank.

Proof: Let p: Fisher is unemployed.

q: Fisher is on welfare.

s: Fisher has money in the bank.

Then the premises are

$$p$$
$$q$$
$$(\sim p) \to s$$
$$q \to (\sim s)$$

and the conclusion is $(\sim s)$.

Let us assume that $(\sim s)$ is false. Then s is true. But $q \to (\sim s)$ is true and q is true, therefore $(\sim s)$ is true by the law of detachment. We now have a contradiction, $(\sim s)$, both true and false. Therefore our assumption is false and $(\sim s)$ is true.

EXERCISE 5

Prove the following.

1. p

q

$(p \land q) \to (r \lor q)$

$(p \lor q) \to (r \land q)$

$\therefore r$

2. $(p \land q) \to (r \land s)$

$\sim s$

q

$\therefore \sim p$

3. p

   q

   $(p \lor q) \to r$

   $\therefore r$

4. $p \to q$

   $r \to (\sim q)$

   $\therefore p \to (\sim r)$

5. If this is a good course, then it is worth taking. Either math is easy or this course is not worth taking. Math is not easy. Therefore this is not a good course.

6. If I don't save my money, I shall not go to Europe. I shall go to Europe. Therefore I save my money.

7. If Carl is elected class president, then Bill will be elected vice-president. If Bill is elected vice-president, then Betty will not be elected secretary. Therefore if Betty is elected secretary, then Carl will not be elected president.

8. Whenever it is not snowing the temperature is high. When the temperature is high, there cannot be ice on the streets. There is ice on the streets. Therefore it is snowing.

9. Vincent water-skis only in summer. Whenever Blanche is in town Vincent water-skis. It is not summer now. Therefore Blanche is not in town.

10. If Martin is on the dean's list, his father will buy him a car. If Martin gets a car, he will drive to Mexico. If Martin does not visit Monterrey, he will not visit Mexico. Martin is on the dean's list. Therefore Martin will visit Monterrey.

11. John is a thief. Newton is a shoplifter. If John is not a thief, then Carl is guilty of car theft. If Newton is a shoplifter, then Carl is not guilty of car theft. Therefore Carl is not guilty.

12. If mathematics is a fascinating subject, then this book is worth reading. If this book is worth reading, then mathematics is not a fascinating subject. Therefore mathematics is a fascinating subject.

13. If this class is a bore, then the instructor is not interesting. If this class isn't a bore, then the subject is worthwhile. The instructor is interesting. Therefore the subject is worthwhile.

14. Men stare at Kathy. Kathy is attractive. If men do not stare at Kathy, they stare at Frances. If Kathy is attractive, the men do not stare at Frances. Therefore men do not stare at Frances.

PIERRE DE FERMAT (1601–1665), the greatest writer of number theory, was a modest counselor of the parliament of Toulouse who devoted his leisure time to mathematics. Most famous of all Fermat's work is his so-called Last Theorem which states that $x^n + y^n = z^n$ has no solution in integers x, y, and z, all different from zero, when $n \geq 3$. Fermat customarily recorded his thoughts in marginal notes, and about his famous theorem he wrote, "I have a truly remarkable demonstration which this margin is too narrow to contain." Whether Fermat possessed a proof of his last theorem probably never will be known. To date no one has proved the theorem for all values of $n \geq 3$.

Divisibility/2

1. THE INTEGERS

In this chapter we shall consider the set of **integers**

$$\{\ldots, -2, -1, 0, 1, 2, \ldots\}$$

We shall assume that the operations of addition, subtraction, multiplication, and division of these numbers is understood.

The set of integers and their operations have the following **properties**:

I-1. (*Closure Property of Addition*): For all integers a and b, $a + b$ is an integer.

I-2. (*Closure Property of Multiplication*): For all integers a and b, ab is an integer.

I-3. (*Commutative Property of Addition*): For all integers a and b, $a + b = b + a$.

I-4. (*Commutative Property of Multiplication*): For all integers a and b, $ab = ba$.

I-5. (*Associative Property of Addition*): For all integers a, b, and c, $(a + b) + c = a + (b + c)$.

I-6. (*Associative Property of Multiplication*): For all integers a, b, and c, $(ab)c = a(bc)$.

I-7. (*Identity Element for Addition*): There is an integer, 0, called the **additive identity**, such that for all a, $a + 0 = 0 + a = a$.

31

I-8. (*Identity Element for Multiplication*): There is an integer, 1, called the **multiplicative identity**, such that for all a, $a \cdot 1 = 1 \cdot a = a$.

I-9. (*Additive Inverse*): For every integer a, there exists a unique integer $-a$, called the **additive inverse** or **opposite** of a, such that $a + (-a) = 0$.

I-10. (*Distributive Property of Multiplication over Addition*): For all integers a, b, and c, $a(b + c) = ab + ac$.

I-11. (*Property of Zero under Multiplication*): For every integer a, $a \cdot 0 = 0 \cdot a = 0$.

I-12. (*Cancellation Property under Addition*): For all integers a, b, and c, if $a + b = a + c$, then $b = c$.

I-13. (*Cancellation Property under Multiplication*): For all integers a, b, and c, if $ab = ac$, $a \neq 0$, then $b = c$.

I-14. (*Closure Property of Subtraction*): For all integers a and b, $a - b$ is an integer.

Let us consider three subsets of the integers:

$$I_p = \{1, 2, 3, 4, \ldots\}$$
$$I_0 = \{0\}$$
$$I_n = \{-1, -2, -3, \ldots\}$$

The set I_p is called the set of **positive integers** or the **natural numbers**. Set I_p together with the set I_0 is called the set of **nonnegative integers** or the **whole numbers**. Set I_n is called the set of **negative integers**. The integer 0 is neither positive nor negative.

2. MULTIPLES AND DIVISORS

Any integer a is said to be **divisible** by an integer b, $b \neq 0$, if there is a third integer c such that

$$a = bc$$

If both a and b are positive, c is necessarily positive. If a is divisible by b, we say that b is a **divisor** or **factor** of a. We also say that a is a **multiple** of b. We express the fact that a is divisible by b by the symbol

$$b|a$$

For example,

$$2 \mid 6 \text{ since } (2)(3) = 6$$
$$5 \mid 10 \text{ since } (2)(5) = 10$$
$$7 \mid 21u \text{ since } (7)(3u) = 21u$$

We use the symbol

$$b \nmid a$$

to express the negation* of $b \mid a$. Thus

$$2 \nmid 7$$
$$3 \nmid 11$$
$$4 \nmid 21$$

If a is an integer, the multiples of a are all numbers of the form na where n is an integer. For example, the multiples of 2 are the **even integers**

$$\ldots -6, -4, -2, 0, 2, 4, 6, \ldots$$

When we defined divisibility of a by b, we stated that b was not zero. Because of its importance, we shall discuss zero and division.

Division is the inverse operation of multiplication. The division $a \div b$ asks the question: What number multiplied by b gives a? That is, if a and b are integers, then $a \div b$ is the integer c if $a = bc$.

Observe that

$$0 \div 5 = n \rightarrow n \times 5 = 0†$$
$$0 \div (-2) = n \rightarrow n \times (-2) = 0$$
$$0 \div (-3) = n \rightarrow n \times (-3) = 0$$

Since these statements are true if and only if $n = 0$, we conclude that 0 divided by any nonzero integer is 0.

We always say that *division by zero is impossible*. To see why we specify that we can only divide by nonzero integers, let us look at the following:

$$6 \div 0 = n \rightarrow n \times 0 = 6$$
$$(-3) \div 0 = n \rightarrow n \times 0 = -3$$

Since zero times any number is zero, there is no integer n that makes the foregoing statements true. Hence, *division by zero is impossible*. When

* See Chapter 1, Section 6.
† See Chapter 1, Section 7.

we define $a \div b = c$ as $b \times c = a$, we must state $b \neq 0$.

There is one other case to be considered: that is $0 \div 0$. In this case

$$0 \div 0 = n \rightarrow n \times 0 = 0$$

Since zero times any integer is zero, n may be replaced by any integer in $n \times 0 = 0$ to make the statement true. Because of this, we call $0 \div 0$ an **indeterminate symbol**.

We may summarize this discussion by saying that $a \div 0$ is impossible for all integers. That is, *division by zero is impossible*.

We can now prove some important theorems about the divisibility of integers. In the following theorems, although not stated, it is assumed that the divisors are not zero.

THEOREM 2.1: If $a \mid k$ and $a \mid h$, then $a \mid h + k$.

Proof: Since $a \mid k$ and $a \mid h$, we may write

$$k = am$$
$$h = an$$

where m and n are integers. Adding these two equations member by member we have

$$k + h = am + an$$
$$= a(m + n)$$

by the distributive property. Since m and n are integers, their sum $m + n$ is an integer, let us call it N. Then

$$k + h = aN$$

Hence $a \mid k + h$.

THEOREM 2.2: If $a \mid k$ and $a \mid h$, then $a \mid k - h$.

Proof: Since $a \mid k$ and $a \mid h$, we may write

$$k = am$$
$$h = an$$

where m and n are integers. Subtracting these two equations member by member we have

$$k - h = am - an$$
$$= a(m - n)$$

Since m and n are integers, their difference is an integer, call it M. Then

$$k - h = aM$$

Hence

$$a \mid k - h$$

THEOREM 2.3: If $a \mid k$ and $a \mid h$, then $a \mid kh$.

Proof: Since $a \mid k$ and $a \mid h$, we may write

$$k = am$$
$$h = an$$

where m and n are integers. Multiplying these two equations member by member we have

$$kh = (am)(an)$$
$$= a(man)$$

Since a, m, and n are integers, their product is an integer, call it L. Then

$$kh = aL$$

Hence

$$a \mid kh$$

As an illustration of the use of the foregoing theorems, notice that since $2 \mid 4$ and $2 \mid 6$,

$$2 \mid (4 + 6) = 10$$
$$2 \mid (4 - 6) = -2$$
$$2 \mid (4 \cdot 6) = 24$$

When $c = ab$ and $c \neq 0$, clearly both $\pm a$ and $\pm b$ are divisors or factors of c. Thus, since $6 = 2 \cdot 3$, ± 2 and ± 3 are factors of 6. *The integers* $+1$ *and* -1 *are factors of every integer.* In the set of positive integers, $+1$ is the only number that is a factor of all the other positive integers. *Zero is the only integer that is a multiple of every integer.* Every integer n, $n \neq 0$, is divisible by $+n$ and $-n$. Thus all the divisors of the positive integer 6 are ± 1, ± 2, ± 3, and ± 6.

We can now prove:

THEOREM 2.4: If $d \mid c$ and $c \mid a$, then $d \mid a$.

Proof: Since $d \mid c$ and $c \mid a$, we may write

$$c = de$$

$$a = cb$$

where b and e are integers. Using the law of substitution*
we have

$$a = cb = (de)b = d(eb)$$

by the associative property. Since b and e are integers,
their product eb is an integer, call it N. Then

$$a = dN$$

and hence

$$d \mid a$$

For example, since $2 \mid 6$ and $6 \mid 12$, Theorem 2.4 assures us that $2 \mid 12$.

As a consequence of Theorem 2.4, we can say that if $a \mid c$, then c is divisible by all the divisors of a. Thus if an integer n is divisible by 12, it is divisible by all the divisors of 12: ± 1, ± 2, ± 3, ± 4, ± 6, and ± 12.

In writing a positive integer c as the product of two positive factors a and b where neither a nor b is equal to c, the factors a and b cannot both be greater than $\sqrt{c}$. If both a and b were greater than $\sqrt{c}$ we would have their product, ab, greater than $\sqrt{c} \cdot \sqrt{c} = c$, which is impossible since $ab = c$. Hence either a or b must be less than $\sqrt{c}$. We can assume therefore, in looking for a pair of factors a and b of c, that we will have $a \le \sqrt{c}$ and $b \ge \sqrt{c}$. This limits the possible numbers we have to try in determining the factorization of c.

For example, the product of two numbers is 72. Since $\sqrt{72} < 9$, one of a pair of factors of 72 will be less than 9.

$$72 = 1 \times 72$$

$$= 2 \times 36$$

$$= 3 \times 24$$

$$= 4 \times 18$$

$$= 6 \times 12$$

$$= 8 \times 9$$

* See Chapter 1, Section 11.

EXERCISE 1

1. Write 48 as a multiple of each of the following.

    (a) 1   (c) 3   (e) 6
    (b) 2   (d) 4   (f) 8

2. Find all the positive divisors of each of the following.

    (a) 12   (c) 96   (e) 180   (g) 270
    (b) 56   (d) 144   (f) 220   (h) 160

3. Which positive integer is a factor of all other positive integers?
4. Which integer is a multiple of all positive integers?
5. Name all the positive factors of each of the following.

    (a) 2   (c) 11   (e) 41   (g) 83
    (b) 3   (d) 23   (f) 53   (h) 101

6. Which of the following are true statements? If a statement is
   not true for all cases, give a case that makes it false.

    (a) Every positive integer is divisible by 1.
    (b) Zero is a multiple of every positive integer.
    (c) If a is divisible by c and c is divisible by d, then a is divisible
        by d.
    (d) If k divides m, then $k = mp$ where p is an integer.
    (e) If k divides a, b, and c, then k divides abc.
    (f) If k is divisible by 9, then it is divisible by 1, 3, and 6.
    (g) If k is divisible by 72, then it is divisible by 8.
    (h) Every integer is divisible by itself.

7. An integer n is divisible by 24. Name seven other positive divisors
   of n.
8. Write zero as a multiple of each of the following.

    (a) 88   (c) 26   (e) f
    (b) 17   (d) k (f) x

9. Write 144 as the product of two positive factors in as many ways
   as possible.
10. What number is a factor of every even integer?
11. What whole number is a factor of every integer?
12. What number is a multiple of every integer?
13. Every even integer may be represented by $2n$ where n is an integer.
    Every odd integer may be represented by $2n + 1$ where n is an
    integer. Write each of the following as $2n$ or $2n + 1$. For example,
    $18 = 2 \cdot 9$; $27 = 2 \cdot 13 + 1$.

    (a) 17   (c) -53 (e) 6472   (g) -4113
    (b) -86 (d) 98   (f) -346 (h) 7613

14. Every whole number may be written as $3n$, $3n + 1$, or $3n + 2$, where n is an integer. For example,

$$16 = 3 \cdot 5 + 1$$
$$18 = 3 \cdot 6$$
$$5 = 3 \cdot 1 + 2$$

Write the following in one of the forms $3n$, $3n + 1$, or $3n + 2$.

(a) 6   (c) 14   (e) 20   (g) 58
(b) 32   (d) 27   (f) 35   (h) 39

3. DIVISION AND REMAINDERS

Let $b \neq 0$ be an arbitrary nonnegative integer. Every other non-negative integer, a, will be either a multiple of b or will fall between two consecutive multiples of b. That is, there is an integer q such that

$$a = bq$$

or

$$bq < a < b(q + 1)$$

Geometrically, this may be visualized as follows. Starting at 0, the number line may be partitioned into intervals b units in length. Any point with a whole number for a coordinate either lies within one of these intervals or is an endpoint of an interval.

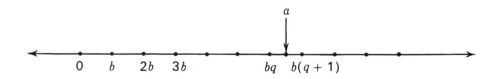

For example, let $a = 18$ and $b = 5$. Since 18 is not a multiple of 5, it must fall between two consecutive multiples of 5. In fact,

$$5 \times 3 < 18 < 5 \times (3 + 1) = 5 \times 4$$

Thus we write

$$a = bq + r$$

where r is one of the numbers, $0, 1, 2, \ldots, (b - 1)$. We call r the **remainder** of a when a is divided by b, whereas q is called the **quotient**. As an example, let us divide 165 by 12. Then

$$165 = (12)(13) + 9$$

The quotient is 13, the remainder is 9. **When a and b are given, q and r are uniquely determined so that each whole number a can be written in the form $a = bq + r$ in only one way where r is one of the numbers 0, 1, 2, 3, $\ldots$, $(b - 1)$.** This property is called the **division algorithm**. We shall accept the division algorithm without proof.

   When each of the whole numbers is divided by 5, the possible remainders are 0, 1, 2, 3, or 4. Hence all the whole numbers may be represented by one of the five forms $5k$, $5k + 1$, $5k + 2$, $5k + 3$, or $5k + 4$, where k is a whole number. The forms of the whole numbers when divided by 5 follow:

$5k$	$5k + 1$	$5k + 2$	$5k + 3$	$5k + 4$
0	1	2	3	4
5	6	7	8	9
10	11	12	13	14
		:		
		:		

Thus 16 is of the form $5k + 1$ because $16 = 5 \cdot 3 + 1$; 27 is of the form $5k + 2$ because $27 = 5 \cdot 5 + 2$; and 25 is of the form $5k$ because $25 = 5 \cdot 5$.

   The division algorithm leads us to many important properties of whole numbers. All whole numbers are either **even** or **odd**. Even numbers are of the form $2k$ and odd numbers are of the form $2k + 1$. When even numbers are squared we have

$$(2k)^2 = 4k^2$$

which is a multiple of 4. When odd numbers are squared we have

$$(2k + 1)^2 = 4k^2 + 4k + 1$$

$$= 4(k^2 + k) + 1$$

by the distributive property. Since k is an integer, $k^2 + k$ is an integer. Let $k^2 + k = M$. Then we have

$$(2k + 1)^2 = 4M + 1$$

We see from this that the square of a whole number is either a multiple of 4 (if it is even), or of the form $4M + 1$, that is, has a remainder of 1 when it is divided by 4 (if it is odd).

We can also show that every square is divisible by 3 or is of the form $3k + 1$. All whole numbers may be represented by one of the forms $3k$, $3k + 1$, or $3k + 2$. If we square each of these in turn we have:

(1)
$$(3k)^2 = 9k^2 = 3(3k^2)$$

Since k is an integer, $3k^2$ is an integer. Let $3k^2 = M$. Then

$$(3k)^2 = 3M.$$

(2)
$$(3k + 1)^2 = 9k^2 + 6k + 1$$

$$= 3(3k^2 + 2k) + 1$$

Since k is an integer, $3k^2 + 2k$ is an integer. Let $3k^2 + 2k = M$. We have

$$(3k + 1)^2 = 3M + 1.$$

(3)
$$(3k + 2)^2 = 9k^2 + 12k + 4$$

$$= 9k^2 + 12k + 3 + 1$$

$$= 3(3k^2 + 4k + 1) + 1$$

Since k is an integer, $3k^2 + 4k + 1$ is an integer. Let $3k^2 + 4k + 1 = M$. We obtain

$$(3k + 2)^2 = 3M + 1$$

From this we see that the square of any whole number is a multiple of 3 or has a remainder 1 when divided by 3 (that is, it is of the form $3k + 1$).

We can also show that the square of an odd number is of the form $8M + 1$. We know from the division algorithm that each whole number is represented by one of the forms $4k$, $4k + 1$, $4k + 2$, or $4k + 3$. Of these, only those numbers of the form $4k + 1$ and $4k + 3$ are odd. Squaring these we see that

$$(4k + 1)^2 = 16k^2 + 8k + 1$$

$$= 8(2k^2 + k) + 1$$

Since k is an integer, so is $2k^2 + k$. Let $2k^2 + k = M$. We obtain

$$(4k + 1)^2 = 8M + 1$$

Similarly

$$(4k + 3)^2 = 16k^2 + 24k + 9$$
$$= 16k^2 + 24k + 8 + 1$$
$$= 8(2k^2 + 3k + 1) + 1$$
$$= 8M + 1.$$

We may state these results as theorems.

THEOREM 2.5: **The square of a whole number is either a multiple of 4 or has a remainder of 1 when divided by 4.**

THEOREM 2.6: **The square of any whole number is either a multiple of 3 or has a remainder of 1 when divided by 3.**

THEOREM 2.7: **The square of any odd number is of the form $8M + 1$.**

We shall now prove that the sum of two even numbers is an even number.

THEOREM 2.8: **The sum of two even numbers is an even number.**

> *Proof:* Every even number may be written in the form $2k$ where k is a whole number. Let $2k$ and $2h$ represent any two even numbers. Then
>
> $$2k + 2h = 2(k + h) \qquad \text{Distributive Property}$$
> $$= 2M \qquad\qquad k + h \text{ is an integer, call it } M$$
>
> But any number of the form $2M$ is an even number. Hence the sum of two even numbers is an even number.

We can also prove

THEOREM 2.9: **The sum of an even number and an odd number is an odd number.**

> *Proof:* Any even number may be written in the form $2k$. Any odd number may be written in the form $2m + 1$.

Then

$$2k + (2m + 1) = (2k + 2m) + 1 \qquad \text{Associative Property}$$

$$= 2(k + m) + 1 \qquad \text{Distributive Property}$$

$$= 2L + 1 \qquad \begin{array}{l} k + m \text{ is an integer,} \\ \text{call it } L. \end{array}$$

But any number of the form $2L + 1$ is odd. Hence the sum of an even number and an odd number is an odd number.

EXERCISE 2

1. Given a and b as follows, find q and r such that $a = bq + r$, $r = 0, 1, 2, \ldots, (b - 1)$.
   (a) $a = 26, b = 8$
   (b) $a = 39, b = 7$
   (c) $a = 126, b = 15$
   (d) $a = 256, b = 27$
   (e) $a = 369, b = 21$
   (f) $a = 1274, b = 97$
   (g) $a = 8, b = 12$
2. If every whole number is divided by 7, the possible remainders are 0, 1, 2, 3, 4, 5, and 6. Hence all whole numbers may be represented by one of seven forms. What are they?
3. Prove: The product of two even numbers is an even number.
4. Prove: The product of two odd numbers is an odd number.
5. Prove: The sum of two odd numbers is an even number.
6. Prove: The fourth power of a number that is not divisible by 5 is of the form $5k + 1$.
7. Prove: If 3 divides a and b (that is, a and b are multiples of 3) then 3 divides $a + b$.
8. Every whole number may be represented by one of the forms:

$$\begin{array}{ll} 12k & 12k + 6 \\ 12k + 1 & 12k + 7 \\ 12k + 2 & 12k + 8 \\ 12k + 3 & 12k + 9 \\ 12k + 4 & 12k + 10 \\ 12k + 5 & 12k + 11 \end{array}$$

(a) Which of the numbers named above are not divisible by 2?
(b) Which of the numbers named above are not divisible by 3?
(c) Which of the numbers named above are not divisible by 2 or 3?
(d) Using the result you found in (c), prove that the square of a number not divisible by 2 or 3 is of the form $12k + 1$.

9. All whole numbers may be represented by one of the forms
$$6k \quad 6k + 1 \quad 6k + 2 \quad 6k + 3 \quad 6k + 4 \quad 6k + 5$$
(a) Which of the numbers named above are odd?
(b) Which of the numbers named above are even?
(c) Which of the numbers named above are divisible by 3?

4. MORE THEOREMS ON DIVISIBILITY

We can now prove more theorems concerning divisibility.

THEOREM 2.10: If $a \neq 0$, then $a \mid 0$ and $a \mid a$.

Proof: Since 0 is a multiple of every integer, we know

$$0 = a \cdot 0$$

But from the definition of divisibility we know

$$a \mid 0$$

Since $a = 1 \cdot a$, we know $a \mid a$ by the definition of divisibility.

THEOREM 2.11: $1 \mid a$ for all a.

Proof: Since $a \cdot 1 = a$, we know from the definition of divisibility that $1 \mid a$.

THEOREM 2.12: If $a \mid b$, then $a \mid bc$ for any c.

Proof: Since $a \mid b$ we know

$$ak = b$$

Multiplying both members of this equation by c we have

$$(ak)c = a(kc) = bc \qquad \text{Associative Property}$$
and hence $a \mid bc$ by the definition of divisibility.

THEOREM 2.13: If $a \mid b$ and $a \mid c$, then $a \mid bx + cy$ for any x and y.

Proof: Since $a \mid b$ and $a \mid c$ we know that there exist integers k and n such that

$$ak = b$$

$$an = c$$

Then

$$bx + cy = (ak)x + (an)y \qquad \text{Law of Substitution}$$

$$= a(kx) + a(ny) \qquad \text{Associative Property}$$

$$= a(kx + ny) \qquad \text{Distributive Property}$$

and

$$a \mid bx + cy$$

EXERCISE 3

In problems 1–5, a, b, c, and d are integers.

1. If $a \mid b$ and $a + b = c$, prove $a \mid c$.
2. If $a \mid c$ and $a + b = c$, prove $a \mid b$.
3. If $d \mid 1$ and d is greater than zero, prove that $d = 1$.
4. If $a \mid b$ and $b \mid a$ ($a \neq 0$, $b \neq 0$), prove $a = b$, or $a = -b$.
5. If $d \mid a$, prove $-d \mid a$.
6. Prove that one of three consecutive integers is divisible by 3.
7. Prove that the sum of three consecutive integers is divisible by 3.
8. Prove that $n(n + 1)(2n + 1)$ is divisible by 6, where n is an integer.
9. Prove that the sum of two even integers is an even integer.
10. Prove that the sum of an even integer and an odd integer is an odd integer.
11. Prove that the product of two odd integers is an odd integer.

5. RULES OF DIVISIBILITY

The numeral of every whole number n with $k + 1$ digits can be written in **expanded form** as

$$n = a \cdot 10^k + b \cdot 10^{k-1} + \cdots + m \cdot 10^3 + h \cdot 10^2 + t \cdot 10 + u$$

where each coefficient $a, b, \ldots, m, h, t$, and u is one of the digits $0, 1, \ldots, 9$. For example,

$$123{,}895 = 1 \cdot 10^5 + 2 \cdot 10^4 + 3 \cdot 10^3 + 8 \cdot 10^2 + 9 \cdot 10 + 5$$

$$3{,}467 = 3 \cdot 10^3 + 4 \cdot 10^2 + 6 \cdot 10 + 7$$

With this in mind, rules for divisibility by 2, 3, 4, 5, 7, 9, and 10 are easily derived.

DIVISIBILITY BY 2

If a number is divisible by 2, then

$$\frac{\cdots + m \cdot 10^3 + h \cdot 10^2 + t \cdot 10 + u}{2}$$

is integral; that is, it is an integer. We may write the above expression as

$$\frac{\cdots + m \cdot 10^3 + h \cdot 10^2 + t \cdot 10}{2} + \frac{u}{2}$$

Observing the first addend in this sum, we see that 10 is a factor of the numerator, hence we may write

$$\frac{10(\cdots + m \cdot 10^2 + h \cdot 10 + t)}{2} + \frac{u}{2}$$

Since $10 = (2)(5)$, $2 \mid 10$. Since $2 \mid 10$, $2 \mid 10(\cdots + m \cdot 10^2 + h \cdot 10 + t)$, by Theorem 2.12. We need concern ourselves only with the units digit to see whether or not a number is divisible by 2. Since the units digit of a numeral must be one of the digits 0, 1, 2, 3, 4, 5, 6, 7, 8, or 9, and of these only 0, 2, 4, 6, and 8 are divisible by 2, it follows that *a number is divisible by 2 if and only if the units digit of its numeral is 0, 2, 4, 6, or 8*. The phrase "if and only if" combines two statements into one. The two statements in the above are: (1) a number is divisible by 2 if the units digit of its numeral is 0, 2, 4, 6, or 8; and (2) if the units digit of the numeral of a number is 0, 2, 4, 6, or 8, then the number is divisible by 2.

DIVISIBILITY BY 5

A similar argument can be offered for divisibility by 5. It is readily seen that $5 \mid 10$ since $10 = (2)(5)$. Hence $5 \mid 10(\cdots + m \cdot 10^2 + h \cdot 10 + t)$. Therefore it is only necessary for 5 to divide the units digit for 5 to divide a number. The only possible units digits are 0, 1, 2, 3, 4, 5, 6, 7, 8, and 9, and of these only 0 and 5 are divisible by 5. Hence *a number is divisible by 5 if and only if the units of digit of its numeral is 0 or 5.*

DIVISIBILITY BY 3

Demonstrating divisibility by 3 requires the rearrangement of the expanded form of the numeral of the number.

$$\cdots + m \cdot 10^3 + h \cdot 10^2 + t \cdot 10 + u$$

$$= \cdots + 1000m + 100h + 10t + u$$
$$= \cdots + (999 + 1)m + (99 + 1)h + (9 + 1)t + u$$
$$= \cdots + 999m + m + 99h + h + 9t + t + u$$

$$= (\cdots + 999m + 99h + 9t) + (\cdots + m + h + t + u)$$

Since the order in addition has no effect on the sum, it is possible to regroup the expanded form of a numeral as shown above.

Now $(\cdots + 999m + 99h + 9t) = 9(\cdots + 111m + 11h + t)$ by the distributive property. Since $3 \mid 9, 3 \mid 9(\cdots + 111m + 11h + t)$. From this we see that if a number is divisible by 3 we must consider $(\cdots + m + h + t + u)$. This expression is the sum of the digits in the numeral of the number. Hence *a number is divisible by 3 if and only if the sum of the digits in its numeral is divisible by 3.*

DIVISIBILITY BY 9

By returning to the expression $\cdots + 999m + 99h + 9t + \cdots + m + h + t + u$ discussed in divisibility by 3, we see that all the terms in $\cdots + 999m + 99h + 9t$ have a factor 9:

$$\cdots + 999m + 99h + 9t = 9(\cdots + 111m + 11h + t)$$

Since $9 \mid 9, 9 \mid 9(\cdots + 111m + 11h + t)$. For a number to be divisible by 9 we restrict our attention to $(\cdots + m + h + t + u)$. Thus, if $9 \mid (\cdots + m + h + t + u)$, that is, *if 9 divides the sum of the digits in the numeral of a number, then the number is divisible by 9.*

DIVISIBILITY BY 10

Divisibility by 10 is a consequence of the rules for divisibility by 2 and 5, since $10 = (2)(5)$. If a number is divisible by 10, it must also be divisible by 2 and by 5. To be divisible by 2 the units digit of the numeral of the number must be 0, 2, 4, 6, or 8. To be divisible by 5 the units digit of the numeral of the number must be 0 or 5. Since 0 is the only number common to these two sets of digits, *a number is divisible by 10 if and only if its numeral ends in 0.*

DIVISIBILITY BY 4

Beginning again with the expanded form of the numeral of a number and regrouping, we have

$$\cdots + m \cdot 10^3 + h \cdot 10^2 + t \cdot 10 + u = 10^2(\cdots + 10m + h) + (10t + u)$$

Since $4 \mid 10^2$, $4 \mid 10^2(\cdots + 10m + h)$. We now restrict our attention to the term $(10t + u)$. If a number is divisible by 4, then $10t + u$ must be divisible by 4. But $10t + u$ is the number named by the tens digit and the units digit of the number in question. Hence *a number is divisible by 4 if and only if the number named by the last two digits in its numeral is divisible by 4.*

DIVISIBILITY BY 7

If a number is divisible by 7, then

$$\frac{\cdots + m \cdot 10^3 + h \cdot 10^2 + t \cdot 10 + u}{7}$$

$$= \frac{\cdots + m \cdot 10^3 + h \cdot 10^2 + t \cdot 10 + u + 20u - 20u}{7}$$

is an integer (note $20u - 20u = 0$). But

$$\frac{10(\cdots + m \cdot 10^2 + h \cdot 10 + t - 2u) + 21u}{7}$$

is an integer only if

$$\frac{\cdots + m \cdot 10^2 + h \cdot 10 + t - 2u}{7}$$

is an integer since $7 \mid 21u$. Notice that

$$\cdots + m \cdot 10^2 + h \cdot 10 + t$$

is the number named by the remaining digits of the numeral of the number with the units digit deleted.

From the above discussion we conclude that *a number is divisible by 7 if the difference obtained by subtracting two times the units digit from the number named by the remaining digits in its numeral is divisible by 7.* As an example let us test whether or not 315 is divisible by 7.

$$315̸$$

$$2 \times 5 = 10$$

$$31 - 10 = 21$$

and 21 is divisible by 7, hence 315 is divisible by 7.

In large numbers divisibility of the difference by 7 will not be apparent by inspection. The process can be repeated until the point is reached where divisibility by 7 is apparent. For example, let us test 132,342 for divisibility by 7.

$$
\begin{array}{r}
132,\ 342̸ \\
2 \times 2 = \quad\quad 4 \\
\hline
132 \quad 30̸ \\
2 \times 0 = \quad\quad 0 \\
\hline
132 \quad 3̸ \\
2 \times 3 = \quad\quad 6 \\
\hline
126̸ \\
2 \times 6 = 12 \\
\hline
0
\end{array}
$$

Zero is divisible by 7, therefore 132,342 is divisible by 7.

EXERCISE 4

1. Which of the following are divisible by 2?
   (a) 168,764   (e) 7,659,430
   (b) 97,842   (f)   578,999
   (c) 43,105   (g)   876,042
   (d) 867,000   (h)   178,401
2. Which of the numbers named in problem 1 are divisible by 3?
3. Which of the numbers named in problem 1 are divisible by 5?
4. Which of the numbers named in problem 1 are divisible by 4?
5. Which of the numbers named in problem 1 are divisible by 9?

6. Which of the following are divisible by 7?
   - (a) 143,724
   - (b) 367,909
   - (c) 876,421
   - (d) 808,669
   - (e) 999,107
   - (f) 777,777

7. If a number is divisible by 6 it must be divisible by 2 and 3 because $2 \cdot 3 = 6$. Which of the following are divisible by 6?
   - (a) 143,724
   - (b) 67,301
   - (c) 82,524
   - (d) 837,102
   - (e) 366,999
   - (f) 421,212

8. Give a rule for divisibility by 15.

9. Give a rule for divisibility by 45.

10. A number is divisible by 12 if and only if it is divisible by 3 and 4. Which of the numbers named in problem 7 are divisible by 12?

LEONHARD EULER (1707–1783) is probably the greatest man of science that Switzerland ever produced. It has been said that "Euler calculated without apparent effort, as men breathe or as eagles sustain themselves in the wind." Euler was certainly one of the most prolific mathematicians in history. It has been estimated that Euler's work would fill one hundred books. Euler was taught mathematics by his father who intended that his son study theology and succeed him as pastor in the village church. Euler obeyed his father and studied theology. Nevertheless, he was sufficiently talented in mathematics to attract the attention of Johann, Daniel, and Nikolaus Bernoulli who became his close friends. Euler could work anywhere under any conditions. He often would compose his memoirs while one of his thirteen children sat on his lap and the others played at his feet. It has been said that he would dash off a mathematical paper in the half hour or so between the first and second calls for dinner.

Prime Numbers/3

1. PRIME AND COMPOSITE NUMBERS

In the previous chapter we discussed some rules for divisibility. The concept of divisibility introduced there indicates the possibility of breaking down or "decomposing" some numbers in terms of others. This leads us to the concept of prime numbers and composite numbers.

Let us consider all the positive integers greater than 1. These integers fall into two classes, prime numbers and composite numbers. An integer $p > 1$ is called a **prime number**, or simply a **prime**, when its only divisors are 1 and p. The first prime number is 2 because its only divisors are 1 and 2. The first few primes are

$$2, 3, 5, 7, 11, 13, 17, 19, 23, 29, 31, 37$$

The only *even* prime is 2 because every other even number is divisible by 2, and hence has at least one divisor other than 1 and itself.

An integer $m > 1$ which has divisors greater than 1 and less than m is called a **composite number**. The first composite number is 4. It has divisors 1, 2, and 4. The first few composite numbers are

$$4, 6, 8, 9, 10, 12, 14, 15, 16$$

The characteristic property of a composite number consists of the possibility of representing it as a product of two factors a and b,

$$m = ab$$

each of which is greater than 1. For example,

$$4 = 2 \times 2 \qquad 27 = 3 \times 9$$
$$6 = 2 \times 3 \qquad 56 = 7 \times 8$$
$$8 = 2 \times 4 \qquad 110 = 10 \times 11$$

Such a representation is impossible for a prime.

We can show that *every composite number is divisible by a prime.* Of all the divisors of a given composite number m, let us select the smallest, p, which is still greater than 1. Now p must be a prime, otherwise it too would have a divisor q greater than 1 and less than p, and q would be a divisor of m. This contradicts the assumption that p, of all the divisors greater than 1 of m, was the smallest.

Every composite number m is divisible by a prime $p \leq \sqrt{m}$. Since m is a composite number it can be represented in the form

$$m = ab$$

where $a > 1$ and $b > 1$. We can suppose $a \leq b$, and then $a \leq \sqrt{m}$. Now if $a > 1$ has a prime factor $p \leq a \leq \sqrt{m}$, p will be a divisor of m.

We now have a practical test to ascertain whether a given number is a prime. It suffices to divide it by the primes less than or equal to its square root. If one divisor succeeds without a remainder, then the number is composite; otherwise it is a prime. Let us test 661. Since $\sqrt{661}$ is between 25 and 26 we need test only primes not exceeding 26; these primes are 2, 3, 5, 7, 11, 13, 17, 19, and 23. Dividing 661 by each of these in turn, we find that none is a divisor of 661, and hence 661 is a prime.

2. THE SIEVE OF ERATOSTHENES

We have just given a method for determining whether or not a number is a prime. This method works well if the number to be tested is not large. When the number to be tested is large, however, the trials become too numerous and burdensome. Innocent as it may seem, the problem of determining whether a given integer is a prime has no general solution.

A simple approach to the problem is called the **Sieve of Eratosthenes**, after the Greek mathematician Eratosthenes (266–194 B.C.). This method consists of writing down all the integers from 2 to the number n, which is to be tested, and sieving out the composite numbers. Two is the smallest prime, and the multiples of 2

$$2 \cdot 2 \quad 2 \cdot 3 \quad 2 \cdot 4 \quad \cdots \quad 2k \quad \cdots$$

occur in the list of integers at intervals of two following 2. Thus we scratch from the list every second number after 2, all of which are composite numbers. Now, 3, the next integer not scratched out, is a prime. Again multiples of 3 occur in the list of integers at intervals of three following 3, so we scratch out every third number after 3. We continue in this fashion. Since every composite number must have a prime factor not exceeding its square root, every composite number in the list must have a prime factor not exceeding $\sqrt{n}$. Thus, by the time we have deleted all multiples of all primes less than or equal to $\sqrt{n}$, we have sieved out all the composite numbers and all those that remain will be primes not exceeding n.

Table 3.1 shows the completed sieve for $n = 100$. Note that since $10^2 = 100$, the process is completed by the time all the multiples of 7 (largest prime less than $\sqrt{100} = 10$) have been struck out. The primes in the table have been circled.

TABLE 3.1. *Sieve of Eratosthenes for* $n = 100$

	(2)	(3)	~~4~~	(5)	~~6~~	(7)	~~8~~	~~9~~	~~10~~
(11)	~~12~~	(13)	~~14~~	~~15~~	~~16~~	(17)	~~18~~	(19)	~~20~~
~~21~~	~~22~~	(23)	~~24~~	~~25~~	~~26~~	~~27~~	~~28~~	(29)	~~30~~
(31)	~~32~~	~~33~~	~~34~~	~~35~~	~~36~~	(37)	~~38~~	~~39~~	~~40~~
(41)	~~42~~	(43)	~~44~~	~~45~~	~~46~~	(47)	~~48~~	~~49~~	~~50~~
~~51~~	~~52~~	(53)	~~54~~	~~55~~	~~56~~	~~57~~	~~58~~	(59)	~~60~~
(61)	~~62~~	~~63~~	~~64~~	~~65~~	~~66~~	(67)	~~68~~	~~69~~	~~70~~
(71)	~~72~~	(73)	~~74~~	~~75~~	~~76~~	~~77~~	~~78~~	(79)	~~80~~
~~81~~	~~82~~	(83)	~~84~~	~~85~~	~~86~~	~~87~~	~~88~~	(89)	~~90~~
~~91~~	~~92~~	~~93~~	~~94~~	~~95~~	~~96~~	(97)	~~98~~	~~99~~	~~100~~

Variations on the sieve method provide the most effective means for computing tables of primes. The best tables of primes are those of D. N. Lehmer which extend beyond ten million.

3. INFINITUDE OF PRIMES

It is quite natural to ask: Is the set of prime numbers an infinite set? This question was answered by the ancient Greeks. Euclid gave a very simple proof that shows there are infinitely many primes.

THEOREM 3.1: There are infinitely many primes.

Proof: Let us assume that there is a finite number of primes and let p be the largest prime. Then, by our assumption, the primes

$$2, 3, 5, 7, \ldots, p$$

taken in their natural order compose the complete set. Now let us form the number

$$N = (2 \times 3 \times 5 \times 7 \times \cdots \times p) + 1$$

If this number N is a prime, it is not contained in the given set since it is greater than p. If N is composite, it has a prime divisor, call it q. When N is divided by any of the primes $2, 3, 5, \ldots, p$, it has a remainder 1 (because it is of the form $p + 1$ where p is the prime considered). Hence q is different from any of the primes $2, 3, \ldots, p$, which were assumed to be all of the known primes. Therefore N must be a prime number larger than any of those given. Hence a prime can always be produced which will not belong to the given set of primes.

EXERCISE 1

1. Using the method of the Sieve of Eratosthenes, find all the primes less than 200.
2. List the even primes.
3. Can the numeral of a prime number greater than 2 end in 0, 2, 4, 6, or 8? Why or why not?
4. Can the numeral of a prime number greater than 5 end in 5? Why or why not?
5. If a number is a prime greater than 5, what are the possible units digits of its numeral?
6. Form $N = (2 \times 3 \times 5 \times \cdots \times p) + 1$ when p is equal to
   (a) 7   (c) 13
   (b) 11   (d) 23
7. Find a prime that divides
   (a) 893   (c) 9999
   (b) 365   (d) 1460
8. Name a prime that is of the form
   (a) $6k + 1$ (b) $6k - 1$
9. Name a prime that is of the form
   (a) $4k + 1$ (b) $4k + 3$
10. Some primes can be written in the form $(1 + n^2)$ for some positive integer n. For example, $5 = 1 + 2^2$, $17 = 1 + 4^2$. Find three more primes of the form $(1 + n^2)$.

11. Some primes can be written in the form $(n^2 - 1)$ for some positive integer n. For example, $3 = 2^2 - 1$. Can you find other primes of this form? Can n be an odd number?

12. Some primes are one more than a power of 2. For example, $5 = 1 + 2^2$. Find three other such primes.

13. Some primes are one less than a power of 2. For example, $3 = 2^2 - 1$. Find three other such primes.

14. Which of the following statements are true? For those statements that are not true, explain why they are not.
    (a) The number 1 is a prime.
    (b) All prime numbers are odd numbers.
    (c) If n is a whole number, then $2n$ is a composite number.
    (d) There are infinitely many primes.
    (e) The only divisors of a prime number p are 1 and p.

4. FUNDAMENTAL THEOREM OF ARITHMETIC

We previously mentioned that every composite number is divisible by a prime. We are now ready to prove the **Fundamental Theorem of Arithmetic**.

THEOREM 3.2: FUNDAMENTAL THEOREM OF ARITHMETIC. Every composite number can be factored uniquely into prime factors.

Proof: The first step is to show that every composite number N is the product of prime factors. We have shown previously (Section 1) that there exists a prime p_1 such that $N = p_1 n_1$. If n_1 is composite, we can find a further prime factor p_2 such that $n_1 = p_2 n_2$. This process can be continued with decreasing numbers, $n_1, n_2, \ldots, n_k$, until n_k is a prime. Now that the existence of a prime factorization has been established, we must prove that it can be done in only one way. Let us assume that there exist two different prime factorizations,

$$N = p_1 p_2 \cdots p_r = q_1 q_2 \cdots q_s$$

where $p_1, p_2, \ldots, p_r, q_1, q_2, \ldots, q_s$ are primes and $r < s$. Since p_1 divides the product of the q's, that is, it divides N, it must divide one of them. Let p_1 divide q_1. Since

q_1 is a prime, it has only two factors, 1 and itself. Since $p_1 > 1$,

$$p_1 = q_1$$

Continuing in this fashion we find

$$p_1 = q_1$$
$$p_2 = q_2$$
$$\vdots$$
$$p_r = q_r$$

Dividing both sides of

$$p_1 p_2 \cdots p_r = q_1 q_2 \cdots q_s$$

by $(p_1 p_2 \dots p_r)$, we have

$$1 = q_{r+1} q_{r+2} \cdots q_s$$

which is impossible since all of the factors $q_{r+1}, q_{r+2}, q_{r+3}, \dots q_s$, are whole numbers greater than 1. Hence $r = s$, and $q_1 = p_1$, $q_2 = p_2, \dots$, and the prime factorization of N is unique.

A composite number may be factored into its prime factors by a method called the **consecutive primes method**. We shall illustrate this method by an example. Let us factor 144 into its prime factors; that is, let us find the **complete factorization** of 144. We begin with the smallest prime, 2, to see whether or not it is a factor of 144. We see by inspection that 144 is divisible by 2:

$$144 = 2 \times 72$$

Since 2 is a factor of 72, we have

$$144 = 2 \times 2 \times 36$$

We see that 2 is a factor of 36, hence

$$144 = 2 \times 2 \times 2 \times 18$$

Again, 2 is a factor of 18, so we have

$$144 = 2 \times 2 \times 2 \times 2 \times 9$$

Since 2 is not a factor of 9, we try the next prime, 3, and we see that

$$144 = 2 \times 2 \times 2 \times 2 \times 3 \times 3 = 2^4 \times 3^2$$

Since all of the factors in the foregoing expression are primes, we have found the **complete factorization** of 144. The essential results of this method may be written in this shortened form:

$$
\begin{array}{r}
2)\overline{144} \\
2)\overline{72} \\
2)\overline{36} \\
2)\overline{18} \\
3)\overline{9} \\
\overline{3}
\end{array}
\qquad
\begin{aligned}
144 &= 2 \times 2 \times 2 \times 2 \times 3 \times 3 \\
&= 2^4 \times 3^2
\end{aligned}
$$

EXERCISE 2

1. Find the prime factors of each of the following.
   (a) 156   (d) 288
   (b) 365   (e) 450
   (c) 404   (f) 1688
2. Find the complete factorization of each of the following.
   (a) 93   (d) 188
   (b) 72   (e) 707
   (c) 118   (f) 3455
3. State the Fundamental Theorem of Arithmetic.
4. What is the largest prime that divides each of the following?
   (a) 4325   (c) 8080
   (b) 589   (d) 252
5. If a number is a prime, what are its divisors?

5. GREATEST COMMON DIVISOR

Let a and b be two positive integers. If a number c divides both a and b, it is called a **common divisor** of a and b. Among the common divisors of a and b there is a greatest one that is divisible by *all* the other common divisors of a and b and is called the **greatest common divisor (g.c.d.)** of a and b. It is usually denoted by the symbol (a, b). If $(a, b) = 1$, we say that a and b are **relatively prime**.

For example, let $a = 8$ and $b = 12$. The divisors of 8 are 1, 2, 4, and 8. The divisors of 12 are 1, 2, 3, 4, 6, and 12. The common divisors of

8 and 12 are 1, 2, and 4. Since 4 is the greatest of the common divisors, it is the greatest common divisor of 8 and 12, thus $(8, 12) = 4$.

In observing the common divisors of 8 and 12, notice that the greatest common divisor is divisible by *all* the other common divisors of the two numbers.

6. EUCLID'S ALGORITHM

We shall now discuss an orderly, systematic process, called **Euclid's Algorithm**, for finding the greatest common divisor of two positive integers.

Euclid's Algorithm is based on the division algorithm: if a and b are positive integers, unique whole numbers q and r can be found such that $a = bq + r$ where $0 \leq r < b$.

We shall demonstrate Euclid's Algorithm by means of examples. Suppose we wish to find $(368, 88)$.

$$\begin{array}{r} 4 \\ 88{\overline{)368}} \\ 352 \\ \hline 16 \end{array}$$

So

$$368 = 88(4) + 16 \qquad \text{or} \qquad 368 - 88(4) = 16$$

Notice that any number that divides 88 and 368 also divides 16 by the distributive property.* This means that we can reduce the problem to one of finding $(88, 16)$.

$$\begin{array}{r} 5 \\ 16{\overline{)88}} \\ 80 \\ \hline 8 \end{array}$$

So

$$88 = 16(5) + 8 \qquad \text{or} \qquad 88 - 16(5) = 8$$

Thus any number that divides both 88 and 16 also divides 8 by the distributive property. So the problem is reduced to finding $(16, 8)$.

$$\begin{array}{r} 2 \\ 8{\overline{)16}} \\ 16 \\ \hline 0 \end{array}$$

* See Chapter 2, Section 1.

So

$$16 = 8(2)$$

That is, 8 divides 16 and there can be no greater divisor of 16 and 8. There-fore, $(16, 8) = 8$. But

$$(8, 16) = (16, 88) = (368, 88)$$

as detailed in the preceding steps. Hence

$$(368, 88) = 8$$

The foregoing work usually is shortened as follows:

$$
\begin{array}{r}
4 \\
88\overline{)368} \\
352 \quad 5 \\
\overline{16\,)88} \\
80 \quad 2 \\
\overline{8\,)16} \\
0
\end{array}
$$

The g.c.d. of the two given numbers is the last *nonzero* remainder in the division process (8 in this case).

Example 1. Find $(564, 27)$.

  Solution:

$$
\begin{array}{r}
20 \\
27\overline{)564} \\
540 \quad 1 \\
\overline{24\,)27} \\
24 \quad 8 \\
\overline{3\,)24} \\
24 \\
\overline{0}
\end{array}
$$

Hence $(564, 27) = 3$.

If $(a, b) = d$, *we can always find integers x and y such that ax + by* $= d$. *If a and b are small, we can find x and y by inspection or by trial and* error. For example, if $a = 5$ and $b = 3$, then $(a, b) = (3, 5) = 1$. We can easily find integers x and y such that $5x + 3y = 1$. For example,

$$5(2) + 3(-3) = 1$$
$$5(5) + 3(-8) = 1$$
$$5(-7) + 3(12) = 1$$

The reader should find other values for x and y.

When a and b are large, it is not always obvious that x and y can be found. In order to find x and y we use Euclid's Algorithm. The computation for $(368, 88) = 8$ by Euclid's Algorithm gives

1. $368 = 88(4) + 16$
2. $88 = 16(5) + 8$
3. $16 = 8(2) + 0$

The italics identify a and b in the division algorithm. We will now reverse these steps to find x and y such that $368x + 88y = 8$. We begin by expressing 8 in terms of 16 and 88 using step 2:

$$8 = 88 - 16(5)$$

We do not simplify the right member of the equality (we would just get 8), but instead leave it intact.

Using step 1 we can write

$$16 = 368 - 88(4)$$

We now substitute this expression for 16 in the expression for 8:

$$8 = 88 - [368 - 88(4)](5)$$
$$8 = 88 - 5[368 - 88(4)]$$

Keeping 88 and 368 (which are b and a respectively) intact, we have

$$8 = 88 - 5(368) + 20(88)$$
$$= (-5)(368) + 88(20 + 1)$$
$$= (-5)(368) + (21)(88)$$

We now have x and y that satisfy

$$368x + 88y = 8$$

We can clearly check our answer:

$$(368)(-5) = -1840$$
$$(88)(21) = 1848$$

and $1848 - 1840 = 8$.

Example 2. Determine (288, 51) and find integers x and y such that

$$288x + 51y = (288, 51)$$

Solution: Using Euclid's Algorithm we have

$$\begin{aligned}
1.\ \ 288 &= 51(5) + 33 \\
2.\ \ \ 51 &= 33(1) + 18 \\
3.\ \ \ 33 &= 18(1) + 15 \\
4.\ \ \ 18 &= 15(1) + 3 \\
5.\ \ \ 15 &= 3(5) + 0
\end{aligned}$$

Thus $(288, 51) = 3$.

Reversing the process to find x and y we have:
Solving step 4 for 3 we have

$$\text{(a)}\quad 3 = 18 - 15(1)$$

Solving step 3 for 15 we have

$$\text{(b)}\quad 15 = 33 - 18(1)$$

Substituting this value for 15 in (a) we have

$$3 = 18 - (1)[33 - 18(1)]$$
$$\text{(c)}\quad 3 = (-1)(33) + (2)(18)$$

Solving step 2 for 18 we have

$$18 = 51 - 33(1)$$

Substituting this value for 18 in (c) we have

$$3 = (-1)(33) + 2[51 - 33(1)]$$
$$\text{(d)}\quad 3 = (2)(51) + (33)(-3)$$

Solving step 1 for 33 we have

$$33 = 288 - (51)(5)$$

Substituting this value for 33 in (d) we have

$$3 = (2)(51) - 3[288 - 51(5)]$$
$$= 288(-3) + 51(17)$$

We now have integers x and y that satisfy

$$288x + 51y = 3$$

We can easily check our answers:

$$288(-3) = -864$$
$$51(17) = 867$$
$$867 + (-864) = 3$$

EXERCISE 3

1. Use Euclid's Algorithm to find the g.c.d. of the following pairs of numbers.
   (a) (56, 24)   (b) (221, 143)
2. Use Euclid's Algorithm to find the g.c.d. of the following pairs of numbers.
   (a) (272, 98)   (b) (139, 49)
3. Use Euclid's Algorithm to find the g.c.d. of the following pairs of numbers.
   (a) (629, 357)   (b) (1472, 1124)
4. Use Euclid's Algorithm to find the g.c.d. of the following pairs of numbers.
   (a) (76084, 63030)   (b) (18416, 17296)
5. Determine integers x and y such that $629x + 357y = (357, 629)$.
6. Determine integers x and y such that $272x + 98y = (272, 98)$.
7. Determine integers x and y such that $139x + 49y = (139, 49)$.
8. Determine integers x and y such that $126x + 98y = (126, 98)$.
9. Determine integers x and y such that $1472x + 1124y = (1472, 1124)$.
10. Determine integers x and y such that $76084x + 63030y = (76084, 63030)$.
11. Which of the following pairs of numbers are relatively prime?
    (a) (4, 6)   (e) (171, 183)
    (b) (8, 9)   (f) (121, 227)
    (c) (50, 63)   (g) (1728, 1512)
    (d) (20, 35)   (h) (6912, 20101)
12. If p and q are primes, what is (p, q)?
13. What is the g.c.d. of 0 and n if n is a positive integer?

7. SOME UNSOLVED PROBLEMS ABOUT PRIMES

Let us look at the first thirty odd primes:

3	37	79
5	41	83
7	43	89
11	47	97
13	53	101
17	59	103
19	61	107
23	67	109
29	71	113
31	73	127

An examination of this list reveals the presence of several pairs of *twin primes*; that is, consecutive primes that differ by 2. Observation of the twin primes in the list leads us to ask the question: How many twin primes are there? Is there an infinity of these primes or is there a largest pair?

The answers to these questions have never been found. Emperical evidence points to the conclusion that there is an infinity of twin primes. It has been shown that there are fifteen pairs of twin primes between 999,999,990,000 and 1,000,000,000,000, and that there are twenty pairs between 1,000,000,000,000 and 1,000,000,010,000. No one to date, however, has succeeded in solving the problem of whether or not there is an infinity of twin primes.

In about 1742, Goldbach, a Russian mathematician, made the following conjecture: **every even number greater than 2 can be written as the sum of two primes.**

Being unable to prove the conjecture, Goldbach sought the assistance of the Swiss mathematician Euler (1707–1783). Euler was unable to prove the conjecture, but was convinced of its truth.

We can easily verify Goldbach's conjecture for small numbers. For example,

$$4 = 2 + 2$$
$$6 = 3 + 3$$
$$8 = 3 + 5$$
$$10 = 3 + 7$$

When the numbers are fairly large, there usually will be numerous representations. For example,

$$
\begin{aligned}
48 &= 5 + 43 \\
&= 7 + 41 \\
&= 11 + 37 \\
&= 17 + 31 \\
&= 19 + 29
\end{aligned}
$$

Goldbach's conjecture has been verified for numbers up to 100,000. In 1931 an unknown Russian mathematician, Schnirelmann, succeeded in proving that every even number can be represented as the sum of not more than 300,000 primes. In 1937 the Russian mathematician Vinogradoff proved that every even number "beyond a certain point" is the sum of four primes. Even though progress has been made toward the solution, Goldbach's conjecture is still unproved.

There are many other unsolved problems about primes, some of which include:

1. There are infinitely many primes $p = 4m + 3$ such that $q = 2p + 1$ is also a prime:

$$
\begin{aligned}
p &= 4 \cdot 2 + 3 = 11 \\
q &= 2 \cdot 11 + 1 = 23
\end{aligned}
$$

2. There are infinitely many primes of the form $n^2 - 2$:

$$
\begin{aligned}
2^2 - 2 &= 2 \\
3^2 - 2 &= 7
\end{aligned}
$$

3. There are infinitely many numbers n for which $n^2 - 2$ is twice a prime:

$$
\begin{aligned}
4^2 - 2 &= 2 \cdot 7 \\
6^2 - 2 &= 2 \cdot 17
\end{aligned}
$$

4. There are infinitely many primes p such that $q = 2p + 1$ is also a prime:

$$
\begin{aligned}
p &= 11 & q &= 22 + 1 = 23 \\
p &= 2 & q &= 4 + 1 = 5
\end{aligned}
$$

EXERCISE 4

1. Find six pairs of twin primes between 127 and 233.
2. Write the following as the sum of two primes.

   (a) 20   (e) 110
   (b) 30   (f) 150
   (c) 32   (g) 226
   (d) 50   (h) 200

3. Find six primes of the form $n^2 - 2$.
4. Find three numbers n such that $n^2 - 2$ is twice a prime.
5. Find three primes p such that $q = 2p + 1$ is also a prime.
6. It has been conjectured that every even integer is the difference of two primes in infinitely many ways. For example: $12 = 19 - 7$ or $29 - 17$ or $23 - 11$. Write 28 as the difference of two primes in three ways.
7. It has been conjectured that every even integer is the difference of two *consecutive* primes in an infinitude of ways. For example: $6 = 29 - 23$; $37 - 31$; $53 - 47$; $99929 - 99923$. Write 10 as the difference of two consecutive primes in three ways.
8. Between n^2 and $n^2 - n$ there exists at least one prime. Find at least one prime between n^2 and $n^2 - n$ when n has the following values:

   (a) 2   (c) 5
   (b) 4   (d) 10

9. There are at least four primes between the squares of consecutive primes greater than 3. Find the primes between 5^2 and 7^2.

KARL FRIEDRICH GAUSS (1777–1855) was born the son of a day laborer in Germany. It was Gauss who said "Mathematics is the queen of the sciences and the theory of numbers is the queen of mathematics." While still in his teens Gauss constructed a regular polygon of seventeen sides, thus settling a 2000-year-old question. During his university career he conceived the idea of least squares. He introduced the theory of congruences in his Disquisitiones Arithmeticae, his most important work on number theory.

Congruences/4

1. FINITE NUMBER SYSTEMS

A **mathematical system*** is any nonempty set of elements together with one or more operations defined on the elements of the set. Most people are familiar with (1) the system of whole numbers and their operations; (2) the system of rational numbers and their operations; and (3) the system of integers and their operations. All these familiar systems are **infinite systems**, that is, the set of elements in the system is an infinite set. We are now going to study a mathematical system that is a **finite system**, that is, the number of elements in the system is finite.

The particular finite system that we shall study is called a **modular system**. This system results from the use of a numeration system similar to that of the numerals on the face of a clock.

For simplicity, let us examine a modular system by using the seven-hour clock shown in Figure 4.1. This clock resembles a timer. The

figure 4.1

---

* We shall study more mathematical systems in Chapter 6.

elements in this system are 0, 1, 2, 3, 4, 5, and 6. Counting in this system repeats these numbers over and over: 1, 2, 3, 4, 5, 6, 0, 1, 2, 3, 4, 5, 6, 0, 1, 2, 3, 4, 5, 6, 0, 1,....

We are now ready to define the operations in this modular system, called the **modulo-seven system**.

ADDITION

The addition of $2 + 3$ is defined thus: The hand of the clock starts at 0, it turns two spaces in a clockwise direction, and then it turns three more spaces in a clockwise direction. Thus

$$2 + 3 = 5$$

Using this definition of addition, we see

$$3 + 5 = 1$$
$$4 + 3 = 0$$
$$2 + 6 = 1$$
$$5 + 5 = 3$$

Other examples, such as $1 + 2 = 3$, $3 + 3 = 6$, and so on, make it clear why we regard this operation as addition. Results of other combinations, however, do not always agree with ordinary addition. Using this definition, an addition table for the modulo-seven system is shown in Table 4.1. We call this addition "**addition modulo 7**" or "**addition mod 7**." The system is called the **modulo-seven system** because there are seven elements in the system.

TABLE 4.1 *Modulo-seven addition table*

+	0	1	2	3	4	5	6
0	0	1	2	3	4	5	6
1	1	2	3	4	5	6	0
2	2	3	4	5	6	0	1
3	3	4	5	6	0	1	2
4	4	5	6	0	1	2	3
5	5	6	0	1	2	3	4
6	6	0	1	2	3	4	5

Addition modulo 7 has some of the properties of ordinary addition. Notice that in the addition table only the numbers 0, 1, 2, 3, 4, 5, and 6 appear. In other words, if we consider the set of seven numbers 0, 1, 2, 3, 4, 5, and 6 and combine them under the rule of addition mod 7, the sum of any two numbers is again an element of this set. We say that the modulo-seven system is closed under the operation of addition mod 7. This property is called the **closure property of addition.**

If 0 is added to any number in this system, the result is the given number; that is, if a is any element in the system,

$$a + 0 = 0 + a = a$$

The number 0 is the only number in the system which has this property. It is called the **additive identity**.

As we look at the addition table we see that each row and column of the table contains 0 exactly once. This corresponds to another property of addition mod 7. For example, in the fourth row, 0 occurs in the fifth column. This corresponds to the fact that

$$3 + x = 0$$

has the solution $x = 4$ and this is the only solution. Similarly, for any a in the system,

$$a + x = 0$$

has one and only one solution. We express this by saying for any element a in the system, there exists an element called the **additive inverse of a** denoted by $-a$ such that

$$a + (-a) = 0$$

The addition table is symmetric about the diagonal from the upper left corner to the lower right corner. That is, the entry in the ith row and the kth column is the same as the entry in the kth row and the ith column. This means that, for all elements a and b in the system,

$$a + b = b + a$$

We express this fact by saying that addition mod 7 is **commutative**.

There is another property which is not apparent from the table, but which is very important. Notice

$$(1) \quad (2 + 3) + 6 = 5 + 6 = 4$$
$$2 + (3 + 6) = 2 + 2 = 4$$

$$\therefore (2 + 3) + 6 = 2 + (3 + 6)$$
$$(2) \quad (4 + 5) + 6 = 2 + 6 = 1$$
$$4 + (5 + 6) = 4 + 4 = 1$$

$$\therefore (4 + 5) + 6 = 4 + (5 + 6)$$

In general, this is true: if a, b, and c are elements in the system, then

$$(a + b) + c = a + (b + c)$$

This is called the **associative property of addition**.

MULTIPLICATION

In the modulo-seven system we define 4×3 thus: the hand of the clock starts at 0 and makes four turns of three spaces each in a clockwise direction. Thus

$$4 \times 3 = 3 + 3 + 3 + 3$$
$$= 5$$

Using this definition of multiplication mod 7, we see

$$2 \times 3 = 6 \text{ (2 clockwise turns of 3 spaces each)}$$
$$4 \times 5 = 6 \text{ (4 clockwise turns of 5 spaces each)}$$
$$6 \times 6 = 1 \text{ (6 clockwise turns of 6 spaces each)}$$

TABLE 4.2. *Modulo-seven multiplication table*

×	0	1	2	3	4	5	6
0	0	0	0	0	0	0	0
1	0	1	2	3	4	5	6
2	0	2	4	6	1	3	5
3	0	3	6	2	5	1	4
4	0	4	1	5	2	6	3
5	0	5	3	1	6	4	2
6	0	6	5	4	3	2	1

A multiplication table for the modulo-seven system is shown in Table 4.2. Studying the multiplication table for the modulo-seven system, we discover:

1. The set of elements is **closed** under multiplication.
2. Multiplication is **commutative**; that is, for all elements a and b, $a \times b = b \times a$.
3. There is an element, 1, called the **multiplicative identity**, such that for all elements a in the system, $a \times 1 = 1 \times a = a$.
4. Every element a, except 0, has a **multiplicative inverse** denoted by a^{-1} such that $a \times a^{-1} = 1$.
5. Again, it is not obvious from the table but it is true that, for all a, b, and c in the system, $(a \times b) \times c = a \times (b \times c)$. This is the **associative property of multiplication**.

The two operations of addition and multiplication mod 7 are connected by the **distributive property**, which states that if a, b, and c are elements of the system, then

$$a \times (b + c) = (a \times b) + (a \times c)$$

For example,

$$2 \times (3 + 5) = 2 \times 1 = 2$$
$$(2 \times 3) + (2 \times 5) = 6 + 3 = 2$$

$$\therefore 2 \times (3 + 5) = (2 \times 3) + (2 \times 5)$$

EXERCISE 1

1. Use Table 4.1 to find the following.
   (a) $(3 + 4) + 5$ (d) $(4 + 6) + (6 + 3)$
   (b) $(6 + 3) + 5$ (e) $(5 + 2) + (6 + 5)$
   (c) $(5 + 3) + 2$ (f) $(2 + 3) + (6 + 1)$
2. What is the additive inverse of each of the following elements of the modulo-seven system?
   (a) 2   (c) 6
   (b) 3   (d) 1

3. Using Table 4.2, find:
   (a) $(3 \times 2) \times 6$ (d) $(6 \times 5) \times (5 \times 5)$
   (b) $(4 \times 5) \times 2$ (e) $(3 \times 3) \times (4 \times 5)$
   (c) $(6 \times 6) \times 3$ (f) $(6 \times 3) \times (6 \times 2)$
4. Using Tables 4.1 and 4.2, compute the following.
   (a) $(6 \times 4) + (3 \times 2)$ (c) $(3 \times 2) + (6 \times 4)$
   (b) $6 \times (5 + 4)$ (d) $(6 \times 6) + (5 \times 1)$
5. Using Tables 4.1 and 4.2, show that the following are true.
   (a) $3 \times (5 + 4) = (3 \times 5) + (3 \times 4)$
   (b) $6 \times (4 + 6) = (6 \times 4) + (6 \times 6)$
   (c) $3 \times (1 + 5) = (3 \times 1) + (3 \times 5)$
   (d) $6 \times (5 + 2) = (6 \times 5) + (6 \times 2)$
6. What modular system is based on a clock with numerals 0, 1, 2, 3, and 4?
7. Construct an addition table for a modulo-five system.
8. Construct a multiplication table for a modulo-five system.
9. Answer the following with reference to the modulo-five system.
   (a) Show that $(4 + 2) + 3 = 4 + (2 + 3)$.
   (b) Show that $2 \times (4 + 3) = (2 \times 4) + (2 \times 3)$.
   (c) What is the additive inverse of each element in this system?
   (d) What is the multiplicative inverse of every element $a \neq 0$ of the system?
   (e) Solve the equation $2 + x = 1$.
   (f) Solve the equation $3x = 2$.
10. Construct addition and multiplication tables for a modulo-eight system.
11. Answer the following with reference to the modulo-eight system.
   (a) What is the additive inverse of 2?
   (b) What is the multiplicative inverse of 2?
   (c) Do all elements in this system have multiplicative inverses? If not, which do not?
   (d) Solve the equation $4x = 3$.

2. CLASSIFICATION OF NUMBERS

We have studied in the previous section finite number systems called modular systems. We shall now show a connection between elements of these systems and ordinary integers. For example, for the elements 0, 1, 2, 3, 4, 5, and 6, in the modulo-seven system, we could consider a circle with seven equally spaced divisions as shown in Figure 4.2. Each point of division is marked with one of the numbers 0, 1, 2, 3, 4, 5, and 6, inclusive.

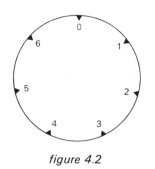

figure 4.2

Now let us mark off on a line the segments equal in length to one-seventh of the circle. Let the circle roll along the line as indicated in Figure 4.3. Because of our choice of units on the line, the point representing 1 on the circle will strike the point representing 1 on the line; the point representing 2 on the circle will strike the point representing 2 on the line; and so on. After the first complete turn of the circle the point representing 0 on the circle will strike the point representing 7 on the line, and so on. If we roll the circle to the left, the point representing 6 on the circle will strike the point representing −1 on the line, and so on.

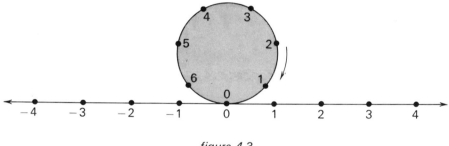

figure 4.3

Thus we see that every point on the line will have a "label" on the circle. Let us see what points on the line correspond to the numbers named on the circle.

0 on the circle represents: ..., −21, −14, −7, 0, 7, 14, ...
1 on the circle represents: ..., −20, −13, −6, 1, 8, 15, ...
2 on the circle represents: ..., −19, −12, −5, 2, 9, 16, ...
3 on the circle represents: ..., −18, −11, −4, 3, 10, 17, ...
4 on the circle represents: ..., −17, −10, −3, 4, 11, 18, ...
5 on the circle represents: ..., −16, −9, −2, 5, 12, 19, ...
6 on the circle represents: ..., −15, −8, −1, 6, 13, 20, ...

We may call this a "**classification**" of the integers mod 7. In the 0-class are all the integers to which the point representing 0 on the circle corresponds; in the 1-class are all the integers to which the point representing 1 on the circle corresponds; and so forth. Notice that:

1. Every integer is in one class.
2. No integer is in two classes.
3. Any two numbers in the same class differ by a multiple of 7 when the circle is marked with the elements in the modulo-seven system.

It is customary to call two numbers in the same class **congruent**. Thus -6 and 8 are congruent in the modulo-seven system. Usually we designate the class in a modulo-m system by the whole number less than m falling in this class. For example, 8 is in the 1-class in the modulo-seven system because there is a multiple of 7 which when subtracted from 8 is 1. That is, there is an integer x such that $8 - 7x = 1$. Similarly, 43 is in the 1-class since $43 - (7)(6) = 1$.

EXERCISE 2

1. Mark a circle into eight parts. Give three integers in each of the eight classes: 0-class, 1-class, 2-class, 3-class, 4-class, 5-class, 6-class, and 7-class.
2. If a circle is marked into eight parts for the modulo-eight system, into which class will each of the following belong?
   (a) 36   (d) 107
   (b) 76   (e) -34
   (c) 85   (f) -100
3. If two numbers belong to the same class for a particular modular system, they are called congruent. Which of the following are congruent to: (1) 8; (2) 4; (3) 6; and (4) 0 for the modulo-nine system?
   (a) 456   (d) 1111
   (b) 67   (e) 999
   (c) 405   (f) 1430
4. Let a circle of m divisions roll on a line as indicated in Figure 4.3. How would you find what number marked on the circle strikes a large number?
5. If a circle is divided into sixteen divisions, how many classes will there be?

3. DEFINITION OF CONGRUENCE

Whenever the question of divisibility of integers by a fixed positive integer m occurs, the concept and notation of **congruence** serves to simplify and clarify the reasoning.

To introduce the concept of congruence let us examine the remainders when integers are divided by 3. We have

$$0 = 0 \cdot 3 + 0 \qquad 6 = 2 \cdot 3 + 0 \qquad -1 = -1 \cdot 3 + 2$$
$$1 = 0 \cdot 3 + 1 \qquad 7 = 2 \cdot 3 + 1 \qquad -2 = -1 \cdot 3 + 1$$
$$2 = 0 \cdot 3 + 2 \qquad 8 = 2 \cdot 3 + 2 \qquad -3 = -1 \cdot 3 + 0$$
$$3 = 1 \cdot 3 + 0 \qquad 9 = 3 \cdot 3 + 0 \qquad -4 = -2 \cdot 3 + 2$$
$$4 = 1 \cdot 3 + 1 \qquad 10 = 3 \cdot 3 + 1 \qquad -5 = -2 \cdot 3 + 1$$
$$5 = 1 \cdot 3 + 2 \qquad 11 = 3 \cdot 3 + 2 \qquad -6 = -2 \cdot 3 + 0$$

$$\vdots \qquad\qquad \vdots$$

Observe that the remainder when any integer is divided by 3 is one of the integers 0, 1, or 2. We say two integers a and b are **congruent modulo 3** when they have the same remainder when divided by 3. Thus $2, 5, 8, 11, \ldots, -1, -4, \ldots$, are all congruent modulo 3. In general, we say that two integers a and b are **congruent modulo m**, where m is a fixed positive integer, if a and b have the same remainder when divided by m. We write

$$a \equiv b \pmod{m}$$

to express the fact that a and b are congruent modulo m. If a and b are integers and a is not congruent to b, modulo m, we write

$$a \not\equiv b \pmod{m}$$

Since $a \equiv b \pmod{m}$ means that a and b have the same remainder when divided by m, the following statements are equivalent to a is congruent to b modulo m.

1. $a = b + mk$ for some integer k.
2. $a - b$ is divisible by m.

To emphasize that $a \equiv b \pmod{m}$ is equivalent to $a = b + mk$ for some integer k, and that $a - b$ is divisible by m, notice

$$5 \equiv 2 \pmod 3 \qquad 5 = 2 + 3 \cdot 1 \qquad 3 \,|\, 5 - 2 = 3$$

$$7 \equiv 1 \pmod 6 \qquad 7 = 1 + 6 \cdot 1 \qquad 6 \,|\, 7 - 1 = 6$$

$$28 \equiv 3 \pmod 5 \qquad 28 = 3 + 5 \cdot 5 \qquad 5 \,|\, 28 - 3 = 25$$

Congruences occur frequently in daily life. For example, the statement that Easter comes on Sunday determines the day modulo 7; the hand on a clock indicates the hour modulo 12; the odometer on a car gives the total mileage traveled modulo 100,000.

EXERCISE 3

1. Which of the following are true statements?
    (a) $6 \equiv 12 \pmod 4$
    (b) $58 \equiv 8 \pmod{10}$
    (c) $127 \equiv 2 \pmod 5$
    (d) $387 \equiv 64 \pmod{17}$
    (e) $-86 \equiv -57 \pmod{17}$
    (f) $-37 \equiv 26 \pmod 2$
2. Give all the values, if any, among the integers 4, 5, 6, 7, and 8 which make the following true
    (a) $a \equiv 0 \pmod 3$
    (b) $a \equiv 2 \pmod 3$
    (c) $a \equiv 1 \pmod 3$
    (d) $a \equiv 16 \pmod{11}$
    (e) $a^2 \equiv 1 \pmod 8$
3. Give all the values, if any, among the integers 4, 5, 6, 7, and 8 which make the following true.
    (a) $2a \equiv 0 \pmod 9$
    (b) $a + 6 \equiv 25 \pmod 7$
    (c) $3a + 5 \equiv 7 \pmod{11}$
    (d) $a^2 \equiv 0 \pmod 6$
    (e) $a^2 - 1 \equiv 4 \pmod 4$
    (f) $a^2 \equiv 2 \pmod 8$
4. For each of the following, give two values of m for which each congruence is true and two for which it is false.
    (a) $16 \equiv 8 \pmod m$
    (b) $19 \equiv 7 \pmod m$
    (c) $-10 \equiv 36 \pmod m$
    (d) $14 \equiv 28 \pmod m$
    (e) $-15 \equiv -9 \pmod m$
5. Find all the integers $0 \le x < 8$ which make the following congruences true.
    (a) $x \equiv 3 \pmod 8$
    (b) $5x \equiv 2 \pmod 8$
    (c) $12x \equiv 16 \pmod 8$
    (d) $36x \equiv 44 \pmod 8$

6. Find all the integers $1 \leq x < 12$ which make the following congruences true.
   (a) $x \equiv 5 \pmod{12}$
   (b) $4x \equiv 6 \pmod{12}$
   (c) $x^2 \equiv 0 \pmod{12}$
7. Find all the integers $0 \leq x < 16$ which make the following congruences true.
   (a) $x \equiv 8 \pmod{16}$
   (b) $x \equiv 9 \pmod{16}$
   (c) $x^2 \equiv 4 \pmod{16}$
8. Find all integers $0 \leq x < 11$ which make the following congruences true.
   (a) $5x \equiv 3 \pmod{11}$
   (b) $x^2 \equiv 1 \pmod{11}$
   (c) $5x \equiv 7 \pmod{11}$
9. Give three positive integers and three negative integers congruent to 3 modulo 8.
10. Give five positive integers and five negative integers congruent to 8 modulo 12.

4. PROPERTIES OF CONGRUENCES

The usefulness of the congruence notation lies in the fact that congruence modulo m has many of the same properties as equality. The most important properties of the relation $a = b$ (equality) are:

1. $a = a$ for all a.
2. If $a = b$, then $b = a$.
3. If $a = b$ and $b = c$, then $a = c$.
4. If $a = b$ and $c = d$, then $a + c = b + d$.
5. If $a = b$ and $c = d$, then $a - c = b - d$.
6. If $a = b$ and $c = d$, then $ac = bd$.

These properties are also true when the relation $a = b$ is replaced by $a \equiv b \pmod{m}$.

1'. $a \equiv a \pmod{m}$ for all a.
2'. If $a \equiv b \pmod{m}$, then $b \equiv a \pmod{m}$.
3'. If $a \equiv b \pmod{m}$ and $b \equiv c \pmod{m}$, then $a \equiv c \pmod{m}$.
4'. If $a \equiv b \pmod{m}$ and $c \equiv d \pmod{m}$, then $a + c \equiv b + d \pmod{m}$.
5'. If $a \equiv b \pmod{m}$ and $c \equiv d \pmod{m}$, then $a - c \equiv b - d \pmod{m}$.
6'. If $a \equiv b \pmod{m}$ and $c \equiv d \pmod{m}$, then $ac \equiv bd \pmod{m}$.

From 4′, 5′, and 6′ we see that congruences with respect to the same modulus may be added, subtracted, and multiplied, member by member. We shall now prove some of these properties of congruence.

THEOREM 4.1: If $a \equiv b$ (mod m) and $c \equiv d$ (mod m), then $a + c \equiv b + d$ (mod m).

Proof: Since $a \equiv b$ (mod m) and $c \equiv d$ (mod m), we may write

$$a = b + mk$$

$$c = d + mh$$

where h and k are integers. Adding these two equations member by member we have

$$a + c = (b + d) + m(k + h)$$

Since h and k are integers, their sum, $h + k$, is an integer, so

$$a + c \equiv b + d \ (\text{mod } m)$$

THEOREM 4.2: If $a \equiv b$ (mod m) and $c \equiv d$ (mod m), then $a - c \equiv b - d$ (mod m).

Proof: Since $a \equiv b$ (mod m) and $c \equiv d$ (mod m), we may write

$$a = b + mk$$

$$c = d + mh$$

where h and k are integers. Subtracting these two equations member by member we have

$$a - c = (b - d) + m(k - h)$$

Since h and k are integers, their difference is an integer, and hence

$$a - c \equiv b - d \ (\text{mod } m)$$

THEOREM 4.3: If $a \equiv b$ (mod m) and $c \equiv d$ (mod m), then $ac \equiv bd$ (mod m).

Proof: Since $a \equiv b$ (mod m) and $c \equiv d$ (mod m), then

$$a = b + mk$$

$$c = d + mh$$

where k and h are integers. Multiplying the two equations member by member we have

$$ac = (b + mk)(d + mh)$$

Simplifying we have

$$ac = bd + bmh + mkd + m^2kh$$

$$= bd + m(bh + kd + mkh)$$

Since $bh + kd + mkh$ is an integer we have

$$ac \equiv bd \text{ (mod } m)$$

We can now prove some additional properties of congruences.

THEOREM 4.4: If $a \equiv b$ (mod m) and n is a positive factor of m, then $a \equiv b$ (mod n).

Proof: Since $a \equiv b$ (mod m) we know that

$$a = b + mk$$

Now n is a factor of m, so $m = nr$ where r is an integer. Substituting we have

$$a = b + (nr)k$$

$$= b + n(rk)$$

and hence

$$a \equiv b \text{ (mod } n)$$

THEOREM 4.5: If $ab \equiv ac$ (mod m), and $(a, m) = 1$, then $b \equiv c$ (mod m).

Proof: Since $ab \equiv ac$ (mod m) we have

$$ab = ac + mk$$

and

$$a(b - c) = mk$$

We see from the preceding statement that either a or $b - c$ is a multiple of m. Since a and m are relatively prime,* a cannot be a multiple of m. Hence $b - c$ is a multiple of m and

$$b \equiv c \pmod{m}$$

THEOREM 4.6: If $(a, m) = d$ and $ab \equiv ac \pmod{m}$, then $b \equiv c \left(\mathbf{mod}\ \dfrac{m}{d}\right)$.

Proof: Since d is a factor of a, let

$$a = a_1 d$$

Since $ab \equiv ac \pmod{m}$

$$ab = ac + mk$$

and

$$a(b - c) = mk$$

Since a and m both have a factor d we can divide both members of the preceding equation by d.

$$a_1(b - c) = \frac{m}{d}k$$

Since $\dfrac{m}{d}$ and a_1 are relatively prime, $b - c$ is a multiple of $\dfrac{m}{d}$ and hence

$$b \equiv c \left(\mathrm{mod}\ \frac{m}{d}\right)$$

Example 1. Since $3 \equiv 12 \pmod{9}$ and $7 \equiv 16 \pmod{9}$,

$$3 + 7 \equiv 12 + 16 \pmod{9} \text{ by Theorem 4.1}$$

$$3 - 7 \equiv 12 - 16 \pmod{9} \text{ by Theorem 4.2}$$

and

$$(3)(7) \equiv (12)(16) \pmod{9} \text{ by Theorem 4.3}$$

* See Chapter 3, Section 5.

Example 2. Since $8 \equiv 26 \pmod 9$ and $3 \mid 9$, by Theorem 4.4,

$$8 \equiv 26 \pmod 3$$

Example 3. Since $(2)(7) \equiv (2)(18) \pmod{11}$ and $(2, 11) = 1$, by Theorem 4.5,

$$7 \equiv 18 \pmod{11}$$

Example 4. Since $(4)(11) \equiv (4)(35) \pmod{12}$ and $(4, 12) = 4$, by Theorem 4.6,

$$11 \equiv 35 \left(\bmod \frac{12}{4}\right)$$

$$11 \equiv 35 \pmod 3$$

Example 5. Since $(16)(5) \equiv (16)(17) \pmod{12}$ and $(16, 12) = 4$, by Theorem 4.6,

$$(4)(5) \equiv (4)(17) \pmod 3$$

5. RESIDUE CLASSES

It is customary when dealing with congruences to use only small, positive integers. If two numbers are congruent for a modulus m, each is called a **residue** of the other modulo m. For example,

$$8 \equiv 2 \pmod 6$$

and hence 8 is called the residue of 2, modulo 6.

Every integer is congruent modulo m to one and only one of the numbers

$$0, 1, 2, 3, \ldots, (m - 1)$$

When an integer a is divided by m, the remainder will be congruent to a, and will be contained in the set

$$0, 1, 2, 3, \ldots, (m - 1)$$

For example, if we divide 362 by 8, the remainder is 2. Hence

$$362 \equiv 2 \pmod 8$$

That every integer is congruent to only one of the numbers $0, 1, \ldots,$ $(m - 1)$ follows from the fact that no two distinct numbers of the set are congruent modulo m since their difference is less than m and, being different from zero, cannot be divisible by m.

If we put together all integers congruent modulo m, they will be distributed into m classes called **residue classes.**

For example, if $m = 5$, we have the integers distributed into five residue classes as follows:

$$
\begin{array}{ccccc}
\vdots & \vdots & \vdots & \vdots & \vdots \\
-10 & -9 & -8 & -7 & -6 \\
-5 & -4 & -3 & -2 & -1 \\
0 & 1 & 2 & 3 & 4 \\
5 & 6 & 7 & 8 & 9 \\
10 & 11 & 12 & 13 & 14 \\
\vdots & \vdots & \vdots & \vdots & \vdots
\end{array}
$$

If from each of the m classes into which all integers are distributed modulo m, we pick one member, the numbers selected

$$r_1, r_2, r_3, \ldots, r_m$$

are representative of the m residue classes and constitute a **complete set of residues** modulo m. The set

$$0, 1, 2, 3, 4, \ldots, (m - 1)$$

is called the set of **least positive residues** modulo m.

For example,

$$0, 1, 2, 3, 4, 5$$

is the set of least positive residues modulo 6, and the set

$$12, 19, 26, 9, 10, 35$$

is a complete set of residues modulo 6 since

$$
\begin{array}{lll}
12 \equiv 0 & 9 \equiv 3 & \\
19 \equiv 1 & 10 \equiv 4 & (\mathrm{mod}\ 6) \\
26 \equiv 2 & 35 \equiv 5 &
\end{array}
$$

If a is an integer relatively prime to m, that is, $(a, m) = 1$, and $r_1, r_2, \ldots, r_m$ is a complete set of residues mod m, then the numbers

$$ar_1, ar_2, ar_3, \ldots, ar_m$$

form another complete set of residues mod m. We can prove that this set is a complete set of residues by showing that no distinct numbers of the set are congruent modulo m. If

$$ar_i \equiv ar_k \pmod{m}$$

then, by Theorem 4.5,

$$r_i \equiv r_k \pmod{m}$$

which is impossible since r_i and r_k belong to different residue classes. In particular,

$$0, a, 2a, 3a, \ldots, (m-1)a$$

represent a complete set of residues modulo m since $(a, m) = 1$.

Because each integer is congruent modulo m to one of the integers

$$0, 1, 2, \ldots, (m-1)$$

it is unnecessary to work with integers greater than m in solving congruences. For example, the congruence

$$27x \equiv 126 \pmod{15}$$

can be reduced to

$$12x \equiv 6 \pmod{15}$$

because $27 \equiv 12 \pmod{15}$ and $126 \equiv 6 \pmod{15}$.

EXERCISE 4

1. Prove: $a \equiv a \pmod{m}$ for all a.
2. Prove: If $a \equiv b \pmod{m}$, then $b \equiv a \pmod{m}$.
3. Prove: If $a \equiv b \pmod{m}$ and $b \equiv c \pmod{m}$, then $a \equiv c \pmod{m}$.
4. Which of the following are complete sets of residues for the given moduli (moduli is the plural of modulus)?

(a) 3, 9, 11 (mod 3)

(b) 8, 14, 25, 35 (mod 4)

(c) 0, 41, 52, 64, 71 (mod 5)

(d) 0, 1, 2^2, 2^3, 2^4, 2^5, 2^6, 2^7, 2^8 (mod 9)

(e) 0, 1, 3, 3^2, 3^3, ..., 3^{15} (mod 17)

5. Find the least positive residue congruent to each of the following.

(a) $(14)(8^2)$ (mod 5)

(b) $(8)(7^2)$ (mod 9)

(c) $45 + 38$ (mod 11)

(d) $16 - 37 + 19$ (mod 8)

6. Give the set of all integers congruent to 3 modulo 9.

7. Find all the integers x such that:

(a) $1 \leq x \leq 50$ and $x \equiv 6$ (mod 11)

(b) $-20 \leq x \leq 20$ and $x \equiv 12$ (mod 17)

(c) $-50 \leq x \leq 50$ and $x \equiv -87$ (mod 18)

8. Replace each congruence by an equivalent one involving numbers from the set of least positive residues.

(a) $56x \equiv 18$ (mod 11)

(b) $37x \equiv -15$ (mod 9)

(c) $139x \equiv -87$ (mod 23)

(d) $278x \equiv 126$ (mod 101)

9. Find the least positive residue modulo 7 of $22 \cdot 51 + 698$.

10. Find a complete set of residues modulo 5 composed entirely of multiples of 9.

11. Show that m consecutive integers form a complete residue system modulo m.

12. Simplify the following by reducing the two sides to relatively prime integers keeping the modulus as large as possible (use Theorem 4.4).

(a) $16x \equiv 20$ (mod 36)   (c) $24x \equiv 42$ (mod 18)

(b) $18x \equiv 26$ (mod 8)   (d) $36x \equiv 16$ (mod 24)

6. GEOMETRIC REPRESENTATION OF CONGRUENCE MODULO m

Let us recall that the concept of congruence has a geometric representation similar to the geometric representation of integers. Usually if we wish to represent the integers geometrically we draw a line and mark on it a sequence of equally spaced dots (called points). One centrally located dot is labeled 0. The points to the right of zero are labeled successively $+1, +2, +3, \ldots$. The points to the left of zero are labeled successively $-1, -2, -3, \ldots$. The labeling is continued indefinitely in both directions. In this way we can find a point on the line corresponding to each integer as shown in Figure 4.4.

$$-6 \quad -5 \quad -4 \quad -3 \quad -2 \quad -1 \quad 0 \quad +1 \quad +2 \quad +3 \quad +4 \quad +5 \quad +6 \quad +7$$

figure 4.4

When we are dealing with integers modulo m, any two congruent numbers are considered the same as far as their behavior on division by a positive integer a is concerned since they leave the same remainder. In order to show this geometrically, we use a circle divided into m equal parts. Any integer when divided by m has as a remainder one of the m numbers

$$0, 1, 2, \ldots, (m - 1)$$

These numbers are placed at equal intervals on the circle. Every integer is congruent modulo m to one of these numbers and hence is represented geometrically by one of these points. Two numbers are congruent modulo m if they are represented by the same point. Since all numbers of the same residue class are congruent, each point on the circle represents a residue class. Figure 4.5 is drawn for the case $m = 8$.

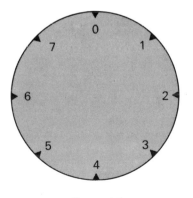

figure 4.5

7. SOLUTION OF CONGRUENCES

We now turn to the problem of solving congruences involving a single unknown x. In general we will have a congruence of the form

$$F(x) \equiv 0 \ (\text{mod } m)$$

where $F(x)$ (read: F of x) is an **integral polynomial** in x; that is, $F(x)$ is an expression built up from the integers and x by addition, subtraction, and multiplication. For example,

$$x^3 - 3x^2 - 4 \equiv 0 \;(\mathrm{mod}\; 9)$$

$$5x^2 + 2x + 3 \equiv 0 \;(\mathrm{mod}\; 7)$$

A congruence such as

$$x^3 + 5x^2 \equiv 3 \;(\mathrm{mod}\; 8)$$

may be handled in the form

$$x^3 + 5x^2 - 3 \equiv 0 \;(\mathrm{mod}\; 8)$$

By a **solution** of a congruence, we mean an integer a such that when x is replaced by a, a true congruence results. Thus 2 is a solution of

$$x^2 + 2x + 1 \equiv 0 \;(\mathrm{mod}\; 9)$$

since

$$2^2 + 2(2) + 1 \equiv 0 \;(\mathrm{mod}\; 9)$$

but 3 is not a solution since

$$3^2 + 2(3) + 1 \equiv 16 \equiv 7 \not\equiv 0 \;(\mathrm{mod}\; 9)$$

Because each integer is congruent to one of the integers

$$0, 1, 2, \ldots, (m - 1)$$

it is unnecessary to work with integers greater than m in solving congruences. For example,

$$9x^2 + 7x + 14 \equiv 0 \;(\mathrm{mod}\; 6)$$

is equivalent to

$$3x^2 + x + 2 \equiv 0 \;(\mathrm{mod}\; 6)$$

because $9 \equiv 3 \;(\mathrm{mod}\; 6)$, $7 \equiv 1 \;(\mathrm{mod}\; 6)$, and $14 \equiv 2 \;(\mathrm{mod}\; 6)$.

We will not always use a reduction, however. For example, the solution of $x^2 \equiv 7 \;(\mathrm{mod}\; 9)$ can be found by trial and error, that is, by substituting the numbers $0, 1, \ldots, 8$ for x and determining which of these numbers are solutions. But $x^2 \equiv 7 \;(\mathrm{mod}\; 9)$ is equivalent to $x^2 \equiv 16 \;(\mathrm{mod}\; 9)$ since $7 \equiv 16 \;(\mathrm{mod}\; 9)$, and this congruence has two obvious solutions, $x \equiv 4$ and $x \equiv -4 \;(\mathrm{mod}\; 9)$.

A second simplification of our problem is most important in the solution of congruences. That is: *if a is a solution of the congruence* $F(x) \equiv 0 \pmod{m}$, *then all the numbers in the set*

$$\ldots, a - 2m, a - m, a, a + m, a + 2m, \ldots$$

are also solutions. Hence, in looking for solutions of congruences, we should look only for solutions among the numbers

$$0, 1, \ldots, (m - 1)$$

because all other integers are congruent to one of these. These solutions can be found by a finite number of calculations. Although this is not the most efficient method of solution, it is one way to solve congruences.

Let us solve some congruences using this method.

Example 1. Solve $x^3 \equiv 3 \pmod{5}$.

   Solution: We test the numbers $0, 1, 2, 3$, and 4 and list those that are solutions.

$$0^3 \not\equiv 3 \pmod{5}$$
$$1^3 \not\equiv 3 \pmod{5}$$
$$2^3 \equiv 8 \equiv 3 \pmod{5}$$
$$3^3 \equiv 27 \equiv 2 \not\equiv 3 \pmod{5}$$
$$4^3 \equiv 64 \equiv 4 \not\equiv 3 \pmod{5}$$

Thus $x \equiv 2 \pmod{5}$ is the only solution. If we are asked for *all* solutions x such that $x^3 \equiv 3 \pmod{5}$, the answer is

$$x = 2 + 5t$$

where t is an arbitrary integer. This gives the set

$$\{\ldots, -13, -8, -3, 2, 7, \ldots\}$$

Example 2. Solve $x^2 + 2x - 1 \equiv 0 \pmod{3}$.

   Solution: We test the numbers $0, 1$, and 2.

$$0^2 + (2)(0) - 1 \not\equiv 0 \pmod{3}$$
$$1^2 + (2)(1) - 1 \not\equiv 0 \pmod{3}$$
$$2^2 + (2)(2) - 1 \not\equiv 0 \pmod{3}$$

Therefore this congruence has no solution.

Example 3. Solve $x^2 \equiv 1 \pmod 8$.

Solution: We test the numbers 0, 1, 2, 3, 4, 5, 6, and 7.

$$0^2 \not\equiv 1 \pmod 8$$
$$1^2 \equiv 1 \pmod 8$$
$$2^2 \equiv 4 \not\equiv 1 \pmod 8$$
$$3^2 \equiv 9 \equiv 1 \pmod 8$$
$$4^2 \equiv 16 \not\equiv 1 \pmod 8$$
$$5^2 \equiv 25 \equiv 1 \pmod 8$$
$$6^2 \equiv 36 \not\equiv 1 \pmod 8$$
$$7^2 \equiv 49 \equiv 1 \pmod 8$$

Therefore the solutions are

$$x \equiv 1 \pmod 8$$
$$x \equiv 3 \pmod 8$$
$$x \equiv 5 \pmod 8$$
$$x \equiv 7 \pmod 8$$

and all solutions are given by

$$x = 1 + 8t$$
$$x = 3 + 8t$$
$$x = 5 + 8t$$
$$x = 7 + 8t$$

where t is an arbitrary integer.

EXERCISE 5

1. Find by trial and error all the solutions $0 \leq x \leq (m - 1)$ of the following.
   (a) $x^2 \equiv 1 \pmod 2$ (d) $x^2 \equiv 5 \pmod 7$
   (b) $x^2 \equiv 1 \pmod 3$ (e) $x^3 \equiv 0 \pmod 8$
   (c) $x^2 \equiv 2 \pmod 3$

2. Find by trial and error all the solutions $0 \leq x \leq (m - 1)$ of the following.
   (a) $x^2 + x - 1 \equiv 0 \pmod 3$
   (b) $2x^2 + 1 \equiv 0 \pmod 5$
   (c) $x^2 - 5x + 6 \equiv 0 \pmod 9$

(d) $x^3 \equiv 1 \pmod 3$
(e) $x^3 \equiv 2 \pmod 3$
3. Find by trial and error all the solutions $0 \le x \le (m-1)$ of the following.
(a) $x^3 \equiv 3 \pmod 7$
(b) $x^2 - 3x + 1 \equiv 0 \pmod 5$
(c) $2x^2 + 4x + 2 \equiv 0 \pmod 6$
(d) $x^3 \equiv 0 \pmod 9$
(e) $5x^2 \equiv 1 \pmod 7$
4. Find by trial and error all the solutions $0 \le x \le (m-1)$ of the following.
(a) $2x^2 \equiv 5 \pmod 6$
(b) $3x \equiv 7 \pmod{12}$
(c) $5x - 2 \equiv 6 \pmod 9$
(d) $7x \equiv 11 \pmod{13}$
(e) $9x \equiv 15 \pmod 6$
5. Draw a geometric representation for the integers modulo 12.
6. Draw a geometric representation for the integers modulo 11.

8. SOLUTION OF LINEAR CONGRUENCES

A **linear congruence** is one that involves only the first power of the variable and constants. Examples of linear congruences are

$$3x \equiv 7 \pmod 9$$
$$x - 1 \equiv 0 \pmod 2$$
$$4x \equiv 6 \pmod 8$$

Linear congruences can always be put in the form

$$ax \equiv b \pmod m$$

where $0 \le a < m$ and $0 \le b < m$. For example,

$$12x - 7 \equiv 6 \pmod 8$$

is equivalent to

$$12x \equiv 13 \pmod 8$$

and

$$4x \equiv 5 \pmod 8$$

since $12 \equiv 4 \pmod 8$ and $13 \equiv 5 \pmod 8$.

If $a \equiv 0 \pmod{m}$ and if $b \equiv 0 \pmod{m}$, every integer is a solution of $ax \equiv b \pmod{m}$; if $a \equiv 0 \pmod{m}$ and $b \not\equiv 0 \pmod{m}$, there is no solution to the congruence. This situation is of no interest and we will always assume that $a \not\equiv 0 \pmod{m}$.

In solving linear congruences we distinguish two cases : $(1) (a, m) = 1$ and $(2) (a, m) = d, d \neq 1$. If $(a, m) = 1$, then the numbers

$$0, a, 2a, 3a, \ldots, (m - 1)a$$

form a complete residue system modulo m. Consequently one and only one of them is in the same residue class as b, that is, one and only one of them is congruent to b modulo m. If this number is ax_0, then x_0 is the unique solution of $ax \equiv b \pmod{m}$, and all other roots are congruent to x_0 modulo m; that is, all solutions are given by

$$x = x_0 + mt$$

where t is an arbitrary integer.

If $(a, m) = d$, then the proposed congruence is impossible (that is, it has no solutions) if b is not divisible by d. This can easily be seen :

$$ax \equiv b \pmod{m}$$

is equivalent to

$$ax = b + mk$$

or

$$ax - mk = b$$

If $d \mid a$ and $d \mid m$, then d must divide b for $ax - mk = b$ to be a true statement.

If $(a, m) = d$, and $d \mid b$, then the given congruence is equivalent to the congruence

$$\frac{a}{d}x \equiv \frac{b}{d}\left(\text{mod } \frac{m}{d}\right)$$

in which $\dfrac{a}{d}$ and $\dfrac{m}{d}$ are relatively prime numbers.* Hence this congruence has a unique solution x_0, modulo $\dfrac{m}{d}$. That is, all its roots as well as the roots of congruence

$$ax \equiv b \pmod{m}$$

* By Theorem 4.4.

are

$$x = x_0 + \frac{m}{d}t$$

where t is an arbitrary integer. Taking $t = 0, 1, \ldots, (d - 1)$, we get exactly d distinct modulo m roots,

$$x_0, x_0 + \frac{m}{d}, x_0 + 2\frac{m}{d}, \ldots, x_0 + (d - 1)\frac{m}{d}$$

for two values of t congruent modulo d lead to two values of x which are congruent modulo m, while the exhibited d numbers belong to different residue classes modulo m. We conclude, thereby, that

$$ax \equiv b \pmod{m}$$

is possible if and only if $(a, m) = d$ divides b, and in this case has exactly d roots.

Finding solutions of linear congruences, that is, finding solutions of congruences of the form

(1) $$ax \equiv b \pmod{m}$$

is equivalent to solving the equation

$$ax = b + mk$$

which is equivalent to

$$ax - mk = b$$

Since $(a, m) = d$ and $d \mid b$ (d must divide b for the congruence to have a solution), we can write

$$a = a_1 d, \qquad b = b_1 d, \qquad m = m_1 d$$

We know by Theorem 4.6 that congruence (1) is equivalent to

(2) $$a_1 x \equiv b_1 \pmod{m_1}$$

and $(a_1, m_1) = 1$. Congruence (2) has only one solution since $(a_1, m_1) = 1$.

Congruence (2) may be written

$$a_1 x = b_1 + m_1 k$$

which is equivalent to

(3) $$a_1 x - m_1 k = b$$

Since a_1 and m_1 are relatively prime, that is, $(a_1, m_1) = 1$, we know that we can find integers P and Q such that

(4) $$a_1 Q + m_1 P = 1*$$

If we multiply each member of (4) by b we have

$$a_1(Qb) + m_1(Pb) = b$$

$$a_1(Qb) - m_1(-Pb) = b$$

From this we see that, if we set $x = Qb$ and $k = (-Pb)$ in equation (3), we have a solution. Thus the solution to congruence (2) is

$$x \equiv Qb \pmod{m_1}$$

All the solutions to the given congruence are given by

$$x \equiv Qb + m_1 t \qquad t = 0, 1, 2, 3, \ldots, (d-1)$$

Study the following examples.

Example 1. Solve $3x \equiv 5 \pmod 7$.

Solution: Since $(3, 7) = 1$, this congruence has only one solution. The given congruence is equivalent to the equation

$$3x - 7m = 5$$

Values for x and m can be found by inspection. For example,

$$3(4) + 7(-1) = 5$$

Hence

$$x \equiv 4 \pmod 7$$

All solutions are given by

$$x = 4 + 7t$$

where t is an arbitrary integer.

Example 2. Solve $18x \equiv 30 \pmod{48}$.

Solution: Since $(18, 48) = 6$, and $6 \mid 30$, this congruence has six solutions. By Theorem 4.6 the given congruence is equivalent to the congruence

$$3x \equiv 5 \pmod 8$$

* See Chapter 3, Section 6.

Since $(3, 8) = 1$, this congruence has only one solution. The congruence $3x \equiv 5 \pmod 8$ is equivalent to the equation

$$3x - 8m = 5$$

We know that we can find integers P and Q such that

$$3Q + 8P = 1$$

because if $(a, b) = 1$, we can find integers x and y such that $ax + by = 1$. Using Euclid's Algorithm we have

$$3 = 8(0) + 3$$
$$8 = 3(2) + 2$$
$$3 = 2(1) + 1$$
$$2 = 1(2)$$

Reversing this process we have

$$1 = 3 - 2(1)$$
$$1 = 3 - 1[8 - 3(2)] = 3(3) + 8(-1)$$

Hence $Q = 3$ and hence

$$x \equiv 3 \cdot 5 \equiv 15 \equiv 7 \pmod 8$$

The solutions of the congruence $18x \equiv 30 \pmod{48}$ are

$$x = 7 + 8t \qquad t = 0, 1, 2, \ldots, 5 \pmod{48}$$

or

$$x \equiv 7 \pmod{48}$$
$$x \equiv 15 \pmod{48}$$
$$x \equiv 23 \pmod{48}$$
$$x \equiv 31 \pmod{48}$$
$$x \equiv 39 \pmod{48}$$
$$x \equiv 47 \pmod{48}$$

All solutions are given by

$$x = 7 + 48t \qquad t \text{ an arbitrary integer}$$
$$x = 15 + 48t \qquad t \text{ an arbitrary integer}$$
$$x = 23 + 48t \qquad t \text{ an arbitrary integer}$$
$$x = 31 + 48t \qquad t \text{ an arbitrary integer}$$
$$x = 39 + 48t \qquad t \text{ an arbitrary integer}$$
$$x = 47 + 48t \qquad t \text{ an arbitrary integer}$$

EXERCISE 6

1. Which of the following congruences have solutions?
   (a) $7x \equiv 5 \pmod{19}$
   (b) $2x \equiv 12 \pmod 4$
   (c) $9x \equiv 8 \pmod 3$
   (d) $108x \equiv 72 \pmod{99}$
   (e) $57x \equiv 82 \pmod{103}$
2. How many solutions (if any) do the congruences in problem 1 have?
3. Find all the solutions of the following.
   (a) $3x \equiv 6 \pmod{24}$
   (b) $3x \equiv 5 \pmod{25}$
   (c) $36x \equiv 8 \pmod{102}$
4. Find all the solutions of the following.
   (a) $144x \equiv 216 \pmod{360}$
   (b) $221x \equiv 111 \pmod{360}$
   (c) $20x \equiv 7 \pmod{15}$
5. Find all the solutions of the following.
   (a) $315x \equiv 11 \pmod{501}$
   (b) $360x \equiv 3072 \pmod{96}$
   (c) $75x \equiv 125 \pmod{175}$

9. DIVISIBILITY AND CONGRUENCES

As an example of the use of congruences, we may determine the remainder when successive powers of 10 are divided by a given number. For example,

$$10 \equiv 1 \pmod 9$$

since $10 = 1 \cdot 9 + 1$. Successively multiplying this congruence by itself we obtain

$$10^2 \equiv 10 \cdot 10 \equiv 1 \cdot 1 \equiv 1 \pmod 9$$
$$10^3 \equiv 10 \cdot 10 \cdot 10 \equiv 1 \cdot 1 \cdot 1 \equiv 1 \pmod 9$$
$$10^4 \equiv 10 \cdot 10 \cdot 10 \cdot 10 \equiv 1 \cdot 1 \cdot 1 \cdot 1 \equiv 1 \pmod 9$$

From this we can show that any positive integer

$$z = a_n 10^n + a_{n-1} 10^{n-1} + a_{n-2} 10^{n-2} + \cdots + a_1 10 + a_0$$

expressed in the decimal system of notation, leaves the same remainder on division by 9 as the sum of its digits

$$t = a_n + a_{n-1} + a_{n-2} + \cdots + a_1 + a_0$$

Since $10 \equiv 1 \pmod 9$ and hence $10^n \equiv 1 \pmod 9$ for all n we have

$$z \equiv a_n 10^n + a_{n-1} 10^{n-1} + a_{n-2} 10^{n-2} + \cdots + a_1 10 + a_0$$
$$\equiv a_n + a_{n-1} + a_{n-2} + \cdots + a_1 + a_0 \pmod 9$$

If a number is divisible by 9 it is congruent to 0 modulo 9, hence, if z is divisible by 9,

$$z \equiv a_n + a_{n-1} + a_{n-2} + \cdots + a_1 + a_0 \equiv 0 \pmod 9$$

In other words, a number is divisible by 9 if the sum of the digits in its numeral is divisible by 9.

Using congruences we can show that a number is divisible by 4 if the number named by the last two digits in its numeral is divisible by 4.

Let us look at the powers of 10:

$$10 \equiv 2 \pmod 4$$
$$10^2 \equiv 0 \pmod 4$$
$$10^3 \equiv 0 \pmod 4$$
$$\vdots$$
$$10^n \equiv 0 \pmod 4$$

Now if a number

$$z = a_n 10^n + a_{n-1} 10^{n-1} + \cdots + a_1 10 + a_0$$

is divisible by 4, it must be congruent to 0 modulo 4. Hence

$$z \equiv a_n 10^n + a_{n-1} 10^{n-1} + \cdots + a_1 10 + a_0$$
$$\equiv a_1 10 + a_0 \equiv 0 \pmod 4$$

But $a_1 10 + a_0$ is the number named by the last two digits in the numeral of the number.

EXERCISE 7

1. Using congruences find a rule for divisibility by 3.
2. Using congruences find a rule for divisibility by 5.
3. Using congruences find a rule for divisibility by 10.
4. Using congruences find a rule for divisibility by 8.
5. Using congruences find a rule for divisibility by 2.

JAMES A. GARFIELD (1831–1881), twentieth president of the United States, was born in Ohio the son of a farmer. He served in the Union Army during the Civil War and was promoted to Major General by Lincoln. In 1863 Garfield entered Congress. He was elected to the Senate in 1880, but before taking office he was elected to the presidency. His short administration was unhappy, terminating on July 2, 1881, when he was shot by Charles J. Guiteau. He died on September 19. In 1876, while a member of Congress, Garfield hit upon a solution to the Pythagorean Theorem during a mathematical discussion with other members of Congress. Since that time his proof, published in the New England Journal of Education, *has been called Garfield's Demonstration.*

The Pythagorean Theorem/5

1. THE PYTHAGOREAN THEOREM

The **theorem of Pythagoras** states that (the area of) the square on the hypotenuse of a right triangle is equal to the sum of (the areas of) the squares on the legs. Figure 5.1 shows the truth of this theorem for a right triangle whose legs measure 3 and 4 units and whose hypotenuse measures 5 units.

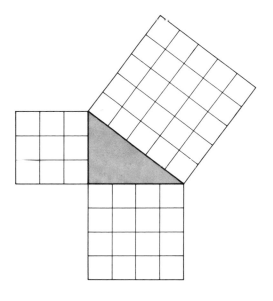

figure 5.1

This relationship between the hypotenuse and the sides of certain right triangles was known in India long before the Christian era, and the Egyptians (2000 B.C.) were aware that the 3-4-5 triangle was right angled. The Chinese used this theorem as early as 1000 B.C., and the Babylonians applied it in 1600 B.C.

Since the time of Pythagoras (540 B.C.) hundreds of different proofs have been given to the Pythagorean theorem, and many descriptive titles have been applied to it. It is probable that the general theorem was due to Pythagoras.

To put the theorem of Pythagoras into modern algebraic form—which Pythagoras could not do—let us denote the measures of the legs by a and b, and the measure of the hypotenuse by c. Then the areas of the squares involved are a^2, b^2, and c^2, and the Pythagorean theorem states

$$a^2 + b^2 = c^2$$

We may ask, "How was Pythagoras led to the discovery of the theorem that bears his name?" History infers that Pythagoras began with the assumption that a triangle with sides 3, 4, and 5 was right angled. But knowing this, how did Pythagoras discover the general theorem? Knowing from the Egyptians that a triangle whose sides have measures of 3, 4, and 5 was a right triangle probably led him to consider whether a similar relation was true of a right triangle whose sides had different measures. The simplest case to investigate was that of a right triangle with legs of equal measure. Such a triangle is called an **isosceles right triangle**. The proof of the theorem in this particular case was from the construction of a figure. Figure 5.2 shows such a construction.

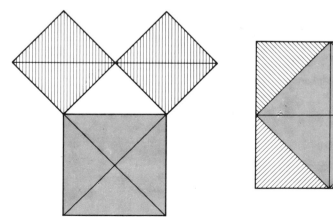

figure 5.2

When it was discovered that the isosceles right triangle had the property in question, Pythagoras was led to establish the property for every right triangle.

To establish a proof of the Pythagorean theorem we start with a right triangle whose legs have measures *a* and *b* and whose hypotenuse has measure *c* (Figure 5.3). Now let us construct the square on the hypotenuse

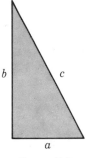

figure 5.3

(Figure 5.4). Note that angle 1 and angle 2 are complementary angles (the sum of their measures is 90°), and that angle 3 is a right angle. Since the sum of the measures of angles 2, 3, and 4 is 180 degrees and the measure

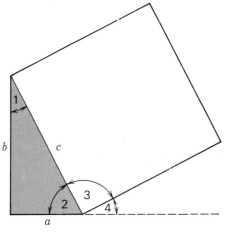

figure 5.4

of angle 3 is 90 degrees, angles 2 and 4 are complementary. We now fit a right triangle congruent to the given right triangle along the base line as shown in Figure 5.5.

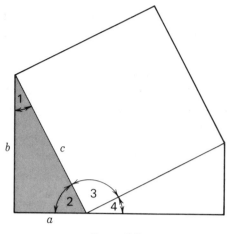

figure 5.5

In a similar manner, two other triangles congruent to the given triangle may be fitted at strategic spots, completing a square (Figure 5.6) with sides of length $a + b$ and area $(a + b)^2$.

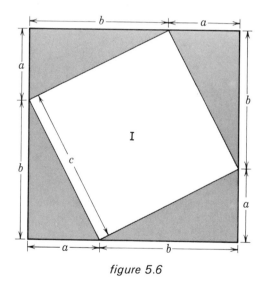

figure 5.6

This square is made up of square I whose area is c^2 and four congruent right triangles each of whose area is $\frac{1}{2}ab$. Hence

$$(a + b)^2 = c^2 + 4(\tfrac{1}{2}ab)$$

Now let us partition this square into appropriate areas as shown in Figure 5.7. This square is made up of square II whose area is b^2 and square III whose area is a^2 and four congruent right triangles each of whose areas is $\frac{1}{2}ab$. Then

$$(a + b)^2 = a^2 + b^2 + 4(\tfrac{1}{2}ab)$$

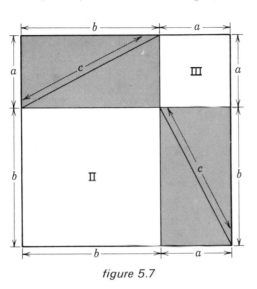

figure 5.7

Since $(a + b)^2 = a^2 + b^2 + 4(\tfrac{1}{2}ab)$ and $(a + b)^2 = c^2 + 4(\tfrac{1}{2}ab)$, we have

$$a^2 + b^2 + 4(\tfrac{1}{2}ab) = c^2 + 4(\tfrac{1}{2}ab)$$

which reduces to

$$a^2 + b^2 = c^2$$

which is the algebraic statement of the Pythagorean theorem.

2. PROOFS OF THE PYTHAGOREAN THEOREM

About 300 B.C., Euclid recorded a proof of the Pythagorean theorem. Euclid's proof of the theorem uses the diagram in which squares are constructed on each side of the right triangle (Figure 5.8). He then proceeded to prove that Area A + Area B = Area C. But since

$$\text{Area } A = a^2$$
$$\text{Area } B = b^2$$
$$\text{Area } C = c^2$$

it is true that

$$a^2 + b^2 = c^2$$

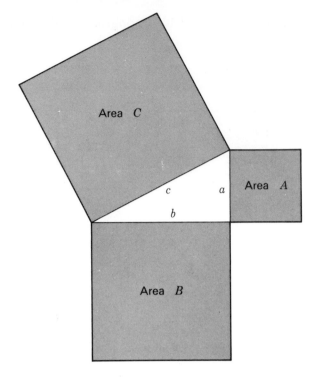

figure 5.8

Another simple proof of the Pythagorean theorem is this: cut out four congruent right triangular regions as shown in Figure 5.9. Place these

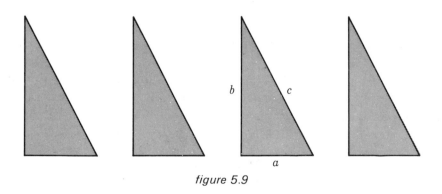

figure 5.9

triangles in the pattern shown in Figure 5.10. Notice that a large square is formed. The measure of its side is c. Its area is c^2. But this area is made up of the area of four right triangles plus the area of the center square. The center square has area $(b - a)^2$. Each right triangle has area $\frac{1}{2}ab$. Hence

$$
\begin{aligned}
c^2 &= (b - a)^2 + 4(\tfrac{1}{2}ab) \\
&= b^2 - 2ab + a^2 + 2ab \\
&= b^2 + a^2
\end{aligned}
$$

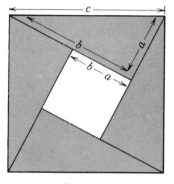

figure 5.10

We shall demonstrate one more proof of the Pythagorean theorem. We start with the right triangle ABC. Now we drop a perpendicular, $\overline{CD}$, from the vertex of the right angle to the hypotenuse. Two pairs of similar

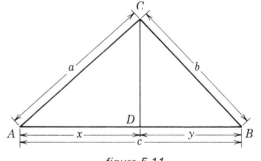

figure 5.11

triangles, $\triangle CDB$ and $\triangle ACB$, and $\triangle ADC$ and $\triangle ACB$, are formed. Triangle CDB and triangle ACB are similar (have the same shape) because both contain $\angle ABC$ and both contain right angles. Since corresponding sides of similar triangles are proportional, we have

$$
\frac{y}{b} = \frac{b}{c} \qquad \text{and} \qquad \frac{x}{a} = \frac{a}{c}
$$

We also know that $x + y = c$. Solving $\dfrac{x}{a} = \dfrac{a}{c}$ and $\dfrac{y}{b} = \dfrac{b}{c}$ for x and y respectively, we have

$$x = \frac{a^2}{c}$$

$$y = \frac{b^2}{c}$$

We know that $x + y = c$, which in turn is equal to

$$\frac{a^2}{c} + \frac{b^2}{c}$$

and we have

$$\frac{a^2}{c} + \frac{b^2}{c} = c$$

which reduces to

$$a^2 + b^2 = c^2$$

EXERCISE 1

1. Which of the following sets of numbers could be measures of the sides of right triangles?

   (a) 10, 24, 26   (e) 1.5, 3.6, 3.9
   (b) 8, 14, 17   (f) $1\frac{1}{2}$, 2, $2\frac{1}{2}$
   (c) 7, 24, 25   (g) $4\frac{1}{4}$, $2\frac{1}{2}$, $5\frac{1}{4}$
   (d) 9, 40, 41   (h) $3\frac{1}{2}$, $4\frac{1}{2}$, $7\frac{1}{2}$

2. Show by the following demonstration that the Pythagorean theorem is intuitively true.

   (a) Prepare seven pieces of cardboard with the measurements shown in the figure below. Label each piece as shown.

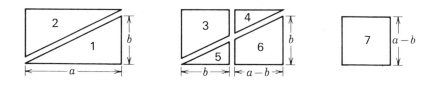

(b) Fit all the pieces together to make a square whose side is *c*.

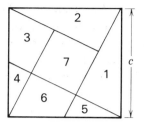

(c) Fit pieces 1, 2, 4, 6, and 7 together to form a square whose side is *a*.

(d) Fit pieces 3 and 5 together to form a square whose side is *b*.

Explain how these steps give evidence of the truth of the Pythagorean theorem.

3. In the rectangular solid below, $AE = 1$, $AB = 2$, $AD = 2$. Find AH.

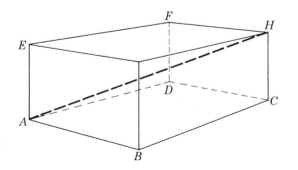

4. If the diagonal of a square is 15 inches, how long is each side of the square?

5. If the diagonal of a rectangle is 25 units and the length is twice the width, what are the dimensions of the rectangle?

6. A man walks 8 miles due north and then 2 miles due east. How far is he from his starting point?

7. A car travels 12 miles west and then 18 miles north. How far is it from its starting point?

8. In the rectangular solid indicated below, find the lengths of $\overline{AC}$ and $\overline{AD}$.

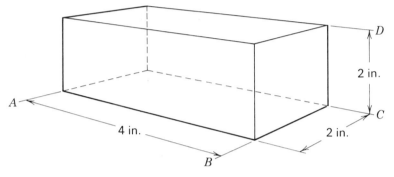

9. The lengths of the legs of a right triangle are 15 and 8. Find the length of the hypotenuse.

10. $\triangle ABC$ has an obtuse angle, $\angle B$. $AB = 6, BC = 14,$ and $AC = 18$. Find the length of the altitude, h, to $\overline{AB}$.

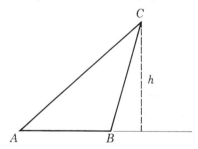

11. In right triangle ABC, $AB = 10$, $CB = 6$. $\overline{EF}$ is perpendicular to and bisects $\overline{AC}$. $AE = 5$. Find ED.

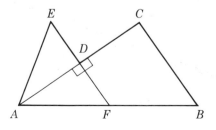

12. A proof of the Pythagorean theorem making use of the figure below was discovered by General James A. Garfield several years before he became President of the United States. It appeared about 1876 in the *New England Journal of Education.*

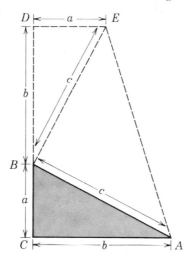

Prove that $a^2 + b^2 = c^2$ by stating algebraically that the area of the trapezoid equals the sum of the areas of the three triangles *ABC*, *BDE*, and *EBA*. The area of a trapezoid is found by taking one-half the product of the altitude and the sum of the bases. (How do you know $\angle EBA$ is a right angle?)

13. It is thought that the Babylonians arrived at the knowledge of the Pythagorean theorem by counting tiles in a pattern similar to the one shown below.

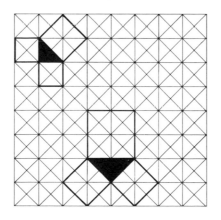

Check the truth of the theorem by counting tiles in the squares drawn on the sides of the shaded triangles.

14. A telephone pole is steadied by three guy wires. Each wire is to be fastened to the pole at a point 15 feet above the ground and anchored to the ground 8 feet from the base of the pole. How many feet of wire are needed for the three guy wires?

15. A gate is 4 feet wide and 6 feet high. How long is the brace that extends from A to B?

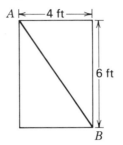

16. A softball diamond is square in shape. The bases are 60 feet apart. How far is it from home plate to second base?

3. PYTHAGOREAN TRIPLES

The algebraic expression of the Pythagorean theorem is

$$x^2 + y^2 = z^2$$

where x and y are the measures of the legs and z is the measure of the hypotenuse.

It is always possible to find triples of real numbers x, y, and z that satisfy the Pythagorean relation $x^2 + y^2 = z^2$. Very interesting problems arise when we attempt to find only those solutions where x, y, and z are positive integers. This problem amounts to finding all triples of positive integers x, y, and z which satisfy the Pythagorean equation $x^2 + y^2 = z^2$.

It suffices to find only those triples that have no common divisor other than 1 (for example, 3, 4, 5 is such a solution; the only common divisor of these numbers is 1), since from such solutions we can find infinitely many other solutions by multiplying each of the three numbers by an arbitrary positive integer M. Thus since 3, 4, 5 is a solution

$$\begin{matrix} 6, & 8, & 10 \\ 9, & 12, & 15 \\ 12, & 16, & 20 \\ & \vdots & \\ 3M, & 4M, & 5M \end{matrix}$$

are all solutions.

Solutions that have no common divisors other than 1 are called **primitive solutions**. In the following discussion we shall be looking for primitive solutions.

Since neither x nor y is zero, z must be greater than either x or y. Furthermore, for integral sides, x cannot be equal to y, for if $x = y = a$, then

$$z^2 = a^2 + a^2$$
$$= 2a^2$$

and

$$z = a\sqrt{2}$$

which is not a positive integer.

Moreover, for a primitive solution, any pair of the numbers x, y, and z must be *relatively prime*.* If, for instance, x and y have a common factor $d \neq 1$, then

$$x = dx_1$$
$$y = dy_1$$

and

$$x^2 + y^2 = (dx_1)^2 + (dy_1)^2$$
$$= d^2x_1^2 + d^2y_1^2$$
$$= d^2(x_1^2 + y_1^2) = z^2$$

and z^2 must be divisible by d^2, and therefore z must be divisible by d. In this case the solution would not be primitive. Similarly, we can prove that x and z cannot have a common factor, nor can y and z.

A consequence of this fact is that if the solution is primitive, x and y cannot both be even. If they were both even, they would have a common factor 2.† Then

$$x = 2x_1 \qquad \text{and} \qquad y = 2y_1$$

Hence

$$x^2 + y^2 = (2x_1)^2 + (2y_1)^2$$
$$= 4x_1^2 + 4y_1^2$$
$$= 4(x_1^2 + y_1^2)$$

Since $x^2 + y^2 = z^2$, we have

$$z^2 = 4(x_1^2 + y_1^2)$$

* See Chapter 3, Section 5.
† See Chapter 2, Section 3.

and z^2 has a factor 4 and z has a factor 2. Thus x, y, and z each would have a factor 2, and the solution would not be primitive.

We can also prove that both x and y cannot be odd if the solution is primitive. Suppose both x and y are odd. Then we can express x as $(2k + 1)$ and y as $(2h + 1)$ where h and k are whole numbers.* Then

$$x^2 = (2k + 1)^2 = 4k^2 + 4k + 1$$
$$y^2 = (2h + 1)^2 = 4h^2 + 4h + 1$$

From this we have

$$
\begin{aligned}
x^2 + y^2 &= 4k^2 + 4k + 1 + 4h^2 + 4h + 1 \\
&= 4(k^2 + k + h^2 + h) + 2 \\
&= 4n + 2
\end{aligned}
$$

Hence z^2 is of the form $4n + 2$. This shows that z^2 is even. But the square of an even number is divisible by 4, and $4n + 2$ is not divisible by 4.*

From the foregoing we see that if we are to find a primitive solution of $x^2 + y^2 = z^2$, x and y must be of different **parity**, that is, one is even and one is odd.

Let us assume that x is even and y is odd. Then z necessarily must be odd since an odd number squared is odd and an even number squared is even and the sum of an even and an odd number is odd.*

We can now write $x^2 + y^2 = z^2$ as

$$z^2 - x^2 = (z - x)(z + x) = y^2$$

Since y is odd, y^2 is odd, and both $(z + x)$ and $(z - x)$ are odd. Furthermore, they are relatively prime, for every common divisor of $(z + x)$ and $(z - x)$ is a common divisor of their sum†

$$(z + x) + (z - x) = 2z$$

and their difference‡

$$(z + x) - (z - x) = 2x$$

and hence is a divisor of the greatest common divisor of $2z$ and $2x$. But x and z are relatively prime, hence $2z$ and $2x$ have only the common divisor 2. Hence, if $(z + x)$ and $(z - x)$ are to have a common divisor other than 1, they could only have common divisor 2. This is impossible because both are odd.

* See Chapter 2, Section 3.
† Theorem 2.1.
‡ Theorem 2.2.

When two numbers are relatively prime, their prime factors are different and their product cannot be a square unless each of them is a square. Since the product of $(z + x)$ and $(z - x)$ is a square, both are squares, so we can write

$$z + x = r^2$$
$$z - x = s^2$$

Since both $(z + x)$ and $(z - x)$ are odd, so are r and s. Since r and s are also odd, both $(r + s)$ and $(r - s)$ are even, and hence divisible by 2. We can write

$$\frac{r + s}{2} = a$$

$$\frac{r - s}{2} = b$$

where a and b are whole numbers. From this we find

$$r = a + b$$
$$s = a - b$$

Then

$$z + x = r^2 = (a + b)^2 = a^2 + 2ab + b^2$$
$$z - x = s^2 = (a - b)^2 = a^2 - 2ab + b^2$$

Solving these equations for z and x, we have

$$z = a^2 + b^2$$
$$x = 2ab$$

But

$$z^2 - x^2 = (a^2 + b^2)^2 - (2ab)^2$$
$$= (a^2 - b^2)^2 = y^2$$

and hence

$$y = a^2 - b^2$$

If both a and b are odd, then $a^2 + b^2$ and $a^2 - b^2$ are even and both y and z would be even, which is impossible since both are odd.

If a and b have a common factor (this would happen if they were both even), then that factor would divide x, y, and z, which is impossible if we are to have a primitive solution.

Hence, for a primitive solution x, y, z, we must have a and b relatively prime and of different parity.

Summarizing the foregoing discussion, we see that x, y, and z is a primitive solution of $x^2 + y^2 = z^2$ if and only if:

1. $x = 2ab, y = a^2 - b^2, z = a^2 + b^2$.
2. a and b are relatively prime.
3. a and b are of different parity.
4. a is greater than b.

Some primitive solutions of $x^2 + y^2 = z^2$ in the smallest values are given in Table 5.1.

TABLE 5.1. *Pythagorean triples*

a	b	x	y	z
2	1	4	3	5
3	2	12	5	13
4	1	8	15	17
4	3	24	17	25

4. SOME PROPERTIES OF PRIMITIVE PYTHAGOREAN TRIANGLES

Any right triangle whose sides have integral measures that are solutions of $x^2 + y^2 = z^2$ is called a **primitive Pythagorean triangle**. We shall discuss a few of the many interesting properties concerning primitive Pythagorean triangles.

It is easy to verify that the measure of one leg of a Pythagorean triangle is always divisible by 3. We know that every square is either a multiple of 3 or of the form $3k + 1$.*

If either x^2 or y^2 is a multiple of 3, then x or y is a multiple of 3 and the property is verified. Suppose neither x^2 nor y^2 is a multiple of 3. Then

$$x^2 = 3k + 1$$

and

$$y^2 = 3h + 1$$

* Theorem 2.6.

From this we have

$$x^2 + y^2 = (3k + 1) + (3h + 1)$$
$$= 3k + 3h + 2$$
$$= 3(k + h) + 2$$

which is of the form $3m + 2$. Then z^2 is of the form $3m + 2$. But a number of the form $3m + 2$ cannot be a square since all squares are of the form $3m$ or $3m + 1$. Hence our assumption is false and either x or y must be a multiple of 3.

We can also prove that the measure of one side of a primitive Pythagorean triangle is divisible by 5.

First we shall prove that every square is either a multiple of 5 or is of the form $5k + 1$ or $5k + 4$. All whole numbers may be written in one of the forms $5k$, $5k + 1$, $5k + 2$, $5k + 3$, or $5k + 4$. Squaring each of these in turn we have

$$(5k)^2 = 25k^2 = 5M$$
$$(5k + 1)^2 = 25k^2 + 10k + 1 = 5M + 1$$
$$(5k + 2)^2 = 25k^2 + 20k + 4 = 5M + 4$$
$$(5k + 3)^2 = 25k^2 + 30k + 9 = 5M + 4$$
$$(5k + 4)^2 - 25k^2 + 40k + 16 = 5M + 1$$

If the measure of one of the legs of a primitive Pythagorean triangle is a multiple of 5, the property is verified. Suppose neither leg is a multiple of 5. We shall show that this implies that the hypotenuse must be a multiple of 5.

If both x^2 and y^2 are of the form $5M + 1$, we have

$$x^2 + y^2 = (5n + 1) + (5p + 1)$$
$$= 5M + 2$$

Hence z^2 is of the form $5M + 2$, which is impossible. From this we know that x^2 and y^2 cannot both be of the form $5M + 1$.

Now, suppose both x^2 and y^2 are of the form $5M + 4$. We have

$$x^2 + y^2 = (5n + 4) + (5p + 4)$$
$$= 5M + 3$$

Hence z^2 is of the form $5M + 3$, which is impossible.

From the foregoing we see that if x is of the form $5M + 1$, then y must be of the form $5M + 4$. Using these conditions we have

$$x^2 + y^2 = (5n + 1) + (5p + 4)$$
$$= 5M$$

and z^2 is divisible by 5, and hence z is divisible by 5.

5. SOLVING PROBLEMS ABOUT PRIMITIVE PYTHAGOREAN TRIANGLES

There are many questions that one might ask in regard to primitive Pythagorean triangles. Examining Table 5.1, we see instances in which the measure of the hypotenuse exceeds the measure of one of the legs by one. The triples 4, 3, 5; 12, 5, 13; and 24, 7, 25 are such examples. We might guess at the answer to this if we study the values for a and b corresponding to x, y, and z.

a	b	x	y	z
2	1	4	3	5
3	2	12	5	13
4	3	24	7	25

In each case we see that $a = b + 1$.

Since the measure of the hypotenuse exceeds the measure of the even leg by 1, we have $z - x = 1$. But $z = a^2 + b^2$, and $x = 2ab$, hence

$$z - x = a^2 + b^2 - 2ab = 1$$

But

$$a^2 + b^2 - 2ab = (a - b)^2$$

hence

$$(a - b)^2 = 1$$

and either $a - b = 1$ or $a - b = -1$. From this either

$$a = b - 1 \qquad \text{or} \qquad a = b + 1$$

Since we have assumed that $a > b$, our condition is given by

$$a = b + 1$$

Now we ask: "If we are given the measure of the odd leg of a primitive Pythagorean triangle, can we find the measures of the other two sides?"

Since

$$y = a^2 - b^2$$
$$= (a - b)(a + b)$$

we factor y into the product of unique, relatively prime factors in all possible ways, and in each case take the larger factor as $a + b = r$, and the smaller factor as $a - b = s$.

For example, given $y = 15$, then

$$15 = 3 \times 5$$
$$15 = 1 \times 15$$

Then $a + b = 5$ and $a - b = 3$, or $a + b = 15$ and $a - b = 1$.

When $a + b = 5$ and $a - b = 3$, we have $a = 4$ and $b = 1$. Hence $x = 8$, $y = 15$, and $z = 17$.

In the case $a + b = 15$ and $a - b = 1$, $a = 8$ and $b = 7$. Then $x = 112$, $y = 15$, and $z = 113$.

If we are given the measure of the even leg of a primitive Pythagorean triangle, can we find the measures of the other two sides?

Since $x = 2ab$ and a and b are of opposite parity, at least one of them is even. Hence x must have a factor 4. We write

$$x = 4M = 2(2M)$$

We now factor $2M$ into a product of two unequal, relatively prime factors of different parity in all possible ways. We then take the larger factor as a and the smaller factor as b and determine y and z.

For example, suppose $x = 24$. Then

$$x = 4 \times 6$$
$$= 2 \times (2 \times 6)$$
$$= 2 \times 12$$

But

$$12 = 1 \times 12 = 3 \times 4$$

Then if $a = 12$ and $b = 1$, $y = 143$ and $z = 145$. If $a = 4$ and $b = 3$, $y = 7$ and $z = 25$.

EXERCISE 2

1. Which of the following sets of numbers are primitive solutions of $x^2 + y^2 = z^2$?

   (a) 1, 2, 3   (f) 5, 12, 13
   (b) 2, 3, 4   (g) 7, 9, 12
   (c) 3, 5, 8   (h) 8, 15, 17
   (d) 4, 5, 6   (i) 9, 12, 15
   (e) 6, 8, 10   (j) 15, 20, 25

2. Find the Pythagorean triples when:

   (a) $a = 2, b = 1$ (e) $a = 11, b = 8$
   (b) $a = 3, b = 2$ (f) $a = 25, b = 16$
   (c) $a = 5, b = 4$ (g) $a = 25, b = 8$
   (d) $a = 7, b = 6$ (h) $a = 18, b = 7$

3. Find all the primitive Pythagorean triples with $z \leq 100$.

4. Find all the primitive Pythagorean triples with $y = 21$.

5. Find all the primitive Pythagorean triples when $y + 2 = z$. That is, y differs from z by 2.

6. Find all the primitive Pythagorean triples when x is equal to:

   (a) 28   (e) 20
   (b) 8   (f) 32
   (c) 12   (g) 36
   (d) 16   (h) 40

7. Find all the primitive Pythagorean triples when y is equal to:

   (a) 7   (d) 27
   (b) 25   (e) 9
   (c) 35   (f) 45

8. Prove that the measure of one leg of a primitive Pythagorean triangle is divisible by 4.

9. Prove that if x, y, and z is a primitive solution of $x^2 + y^2 = z^2$, x and z cannot have a common factor $d \neq 1$.

10. Prove that if x, y, and z is a primitive solution of $x^2 + y^2 = z^2$, y and z cannot have a common factor $d \neq 1$.

6. SQUARE ROOTS FROM THE PYTHAGOREAN TRIANGLES

If the measure of each of the legs of an isosceles right triangle is 1, the measure of the hypotenuse is not a whole number. In fact, it is not even a **rational number** (a rational number is one that may be expressed as

$\frac{a}{b}$, where a and b are integers and $b \neq 0$). We see that if $x = y = 1$, then

$$1^2 + 1^2 = z^2$$
$$1 + 1 = z^2$$
$$2 = z^2$$

and $z = \sqrt{2}$.

There is no whole number that can be multiplied by itself to give 2. Thus the length of the hypotenuse of an isosceles right triangle is **incommensurable**. By commensurable lengths x and z we mean that positive integers m and n can be found such that $mx = nz$. Thus the hypotenuse of an isosceles right triangle is incommensurable with the leg. Another way of saying this is that $\sqrt{2}$ cannot be expressed as the ratio of two integers. Such numbers as $\sqrt{2}$ are called **irrational numbers**.

It is not true that the square root of every whole number is irrational. Thus $\sqrt{4} = 2$, $\sqrt{9} = 3$. However, if the square root of a whole number is not a whole number, it is an irrational number.

There are many ways to find the square root of a whole number. One is by the Pythagorean theorem. If an isosceles right triangle is constructed with 1 for the measure of each of the legs, the measure of the hypotenuse will be $\sqrt{2}$. We now construct another right triangle as shown in Figure 5.12. The hypotenuse of this second triangle has a measure $\sqrt{3}$ since by the Pythagorean theorem it is the square root of $(\sqrt{2})^2 + 1^2$.

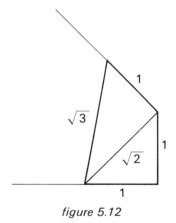

figure 5.12

Continuing in this fashion as shown in Figure 5.13, we can construct right triangles the measures of whose hypotenuses are

$$\sqrt{4}, \sqrt{5}, \sqrt{6}, \ldots$$

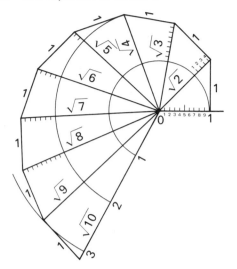

figure 5.13

Continuing this construction indefinitely, we find the general form of the figure is a spiral called the **square root spiral**, with ever-increasing lines extending out from 0 and forming ever-decreasing angles.

It is impractical to find the square root of large numbers by the construction of the square root spiral. If we construct the square root spiral so that we have the square roots of the counting numbers less than or equal to 10, it is possible to use the spiral to find the square roots of other numbers. For example, suppose we wish to find $\sqrt{15}$. The $\sqrt{15}$ can be constructed from $\sqrt{8}$ and $\sqrt{7}$ using Figure 5.13. This construction is shown in Figure 5.14.

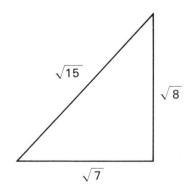

figure 5.14

EXERCISE 3

1. If the measures of the legs of a right triangle are 1 and n, find the measure of the hypotenuse by use of the Pythagorean theorem.

2. If the measures of the legs of a right triangle are $\sqrt{p}$ and $\sqrt{q}$, what is the measure of the hypotenuse?

3. Using the square root spiral construct:

   (a) $\sqrt{13}$ (e) $\sqrt{19}$

   (b) $\sqrt{11}$ (f) $\sqrt{12}$

   (c) $\sqrt{20}$ (g) $\sqrt{16}$

   (d) $\sqrt{17}$ (h) $\sqrt{18}$

4. Triangle ABC is a right triangle. The measures of the legs are a and b; the measure of the hypotenuse is c. How long is the hypotenuse if:

   (a) $a = 6$ and $b = 8$

   (b) $a = 4$ and $b = 6$

5. In an equilateral triangle each side is 15 inches. How long is the altitude?

6. The sides of a triangle are 6 inches, 9 inches, and 11 inches. Is it a right triangle? If it is a right triangle, which side is the hypotenuse?

7. If r and s are lengths of the legs of a right triangle and t is the length of the hypotenuse, show that for any positive integer n, the numbers nr, ns, and nt are also lengths of the sides of a right triangle.

8. How long is the diagonal of a square if its side has measure:

   (a) 6   (d) 13

   (b) 9   (e) $\sqrt{6}$

   (c) $\sqrt{2}$ (f) 17

*EVARISTE GALOIS (1811–1832) invented the term
"group," and his theory, called Galois Theory, is one of the main
sources of finite field theory and of modern abstract algebra.
Galois was educated at the Lycee Louis-de-Grand and Ecole
Normale in Paris. He was a rabid republican and the first shots of
the revolution of 1830 filled Galois with great joy. He was twice
imprisoned for his political views. On May 30, 1832, Galois was
fatally wounded in a duel and died the next day. The night before
the duel took place he wrote a brilliant paper on group theory and
addressed it to his friend Auguste Chevalier who preserved it for
the world. Galois was buried in the common ditch of the South
Cemetery in Paris so that today there is no trace of his grave. All
that remains is his collected work of sixty pages.*

Groups and Fields/6

1. ABSTRACT MATHEMATICAL SYSTEMS

A **mathematical system** is any nonempty set S of elements $\{a, b, c, \ldots\}$, together with one or more **operations** defined on the elements of the set, and a set of **axioms** or **postulates**. In the mathematical systems we shall study, we shall be concerned with operations defined on pairs of elements in the set. Thus the operation will be a binary operation. A **binary operation** on a set S is a rule that assigns to each ordered pair (a, b) of elements of S a *uniquely* defined element of S.

Ordinary addition and multiplication performed on the set of positive integers are binary operations: to each ordered pair of positive integers a and b is assigned unique positive integers c and d, called respectively the **sum** of a and b, and the **product** of a and b, denoted by the symbols

$$(a, b) = a + b = c$$
$$(a, b) = a \times b = d$$

For example, if the operation is addition:

$$(2, 3) = 5$$
$$(3, 6) = 9$$
$$(6, 3) = 9$$

If the operation is multiplication:

$$(2, 3) = 6$$
$$(3, 6) = 18$$
$$(6, 3) = 18$$

In each mathematical system that we shall study, the postulates will be given when the system is structured. Having agreed upon the set, operations, and postulates, we shall investigate some of the propositions that follow from them. We shall call these proved propositions **theorems**, to distinguish them from the assumed propositions or **postulates**.

In any system we shall want an equivalence relation which we shall call *R*. An **equivalence relation** is defined as a relation *R* between the elements of the set *S* which obeys the following rules:

1. *Reflexive Rule:* a R a; that is, the element *a* is related to *a* for every *a* is *S*.
2. *Symmetric Rule:* If a R b, then b R a; that is, if *a* is related to *b*, then this implies that *b* is related to *a* for every *a* and *b* in *S*.
3. *Transitive Rule:* If a R b and b R c, then a R c; that is, if *a* is related to *b*, and *b* is related to *c*, then this implies that *a* is related to *c* for every *a*, *b*, and *c* in *S*.

The most familiar example of an equivalence relation is the ordinary **equality relation**, denoted by the symbol " $=$ ", which relates the elements of a set of numbers: $a = a$ for every *a*, hence the reflexive property is satisfied; if $a = b$, then certainly $b = a$, and the symmetric property is satisfied; if $a = b$ and $b = c$, then $a = c$, and the transitive rule is satisfied. Since the equality relation satisfies the reflexive, symmetric, and transitive rules, it is an equivalence relation. Other examples of equivalence relations include the following.

Example 1. The relation *R*, "lives in the same house as," for the set of all people.

a R a: *a* lives in the same house as *a*.

If a R b, then b R a: If *a* lives in the same house as *b*, then *b* lives in the same house as *a*.

If a R b and b R c, then a R c: if *a* lives in the same house as *b* and *b* lives in the same house as *c*, then *a* lives in the same house as *c*.

Example 2. The relation *R*, "has the same surname (last) as," for the set of all people.

a R a: *a* has the same surname as *a*.

If a R b, then b R a: If *a* has the same surname as *b*, then *b* has the same surname as *a*.

If a R b and b R c, then a R c: if *a* has the same surname as *b*,

and b has the same surname as c, then a has the same surname
as c.

EXERCISE 1

1. Give the unique positive integer associated with each ordered
pair below if the operation is (1) addition, (2) multiplication.
   (a) (116, 349)   (c) (241, 39)
   (b) (74, 968)   (d) (87, 56)
2. Which of the following relations are equivalence relations for the
given sets?
   (a) "Is similar to" for the set of all triangles.
   (b) "Is the brother of" for the set of all men.
   (c) "Has the same initials as" for the set of all people.
3. Which of the following relations are equivalence relations for the
given sets?
   (a) "Is less than" for the set of whole numbers.
   (b) "Is the sister of" for the set of all people.
   (c) "Has the same number of children as" for the set of all married
   couples.
4. Tell which of the three properties of an equivalence relation is
satisfied by each of the relations described in problem 2.
5. Tell which of the three properties of an equivalence relation are
satisfied by each of the relations described in problem 3.
6. Which of the adjectives "reflexive," "symmetric," "transitive" is
applicable to the following relations?
   (a) "Is the father of" for the set of all people.
   (b) "Is East of" for cities in the United States.
   (c) "Lives within three miles of" for the set of all people.
   (d) "Is the wife of" for the set of all people.

2. MATHEMATICAL FIELDS

One of the most familiar mathematical systems is a field. We
define a **field**, F, as a set, S, of elements, $\{a, b, \ldots\}$, with two binary operations, addition and multiplication, denoted respectively by the symbols
$+$ and $\times$, which obey the following postulates.

F-1. (*Closure under Addition*): If a and b are in F, then $a + b$ is in F.

F-2.   (*Commutative under Addition*): If a and b are in F, then $a + b$ $= b + a$.

F-3.   (*Associative under Addition*): If a, b, and c are in F, then $(a + b) + c$ $= a + (b + c)$.

F-4.   (*Additive Identity*): There exists in F an element 0, called the **additive identity**, such that for all a in F, $a + 0 = 0 + a = a$.

F-5.   (*Additive Inverses*): For every element a in F, there exists an element $(-a)$, called the **additive inverse** of a, in F, such that $a + (-a)$ $= (-a) + a = 0$.

F-6.   (*Closure under Multiplication*): If a and b are in F, then $a \times b$ is in F.

F-7.   (*Commutative under Multiplication*): If a and b are in F, then $a \times b = b \times a$.

F-8.   (*Associative under Multiplication*): If a, b, and c are in F, then $(a \times b) \times c = a \times (b \times c)$.

F-9.   (*Multiplicative Identity*): There exists in F an element 1, called the **multiplicative identity**, such that for all a in F, $a \times 1 = 1 \times a = a$.

F-10. (*Multiplicative Inverses*): There exists for every element $a \neq 0$ in F an element a^{-1} in F, called the **multiplicative inverse** of a, such that $a^{-1} \times a = a \times a^{-1} = 1$.

F-11. (*Distributive Property*): If a, b, and c are in F, then $a \times (b + c)$ $= (a \times b) + (a \times c)$.

These postulates fall into three classes:

1. Five postulates describing the behavior of the elements under addition.
2. Five postulates describing the behavior of the elements under multiplication.
3. One postulate connecting addition and multiplication.

The set of whole numbers $\{0, 1, 2, \ldots\}$ and the operations of addition and multiplication do *not* form a field since Postulates F-5 and F-10 are not satisfied. However, the system of whole numbers satisfies all the other postulates.

The system of integers and the operations of addition and multiplication do not form a field since Postulate F-10 is not satisfied. With the exception of $+1$ and -1, no integers have multiplicative inverses.

We shall now consider some common examples of fields.

Example 1. A **rational number** is a number that may be represented by the form $\dfrac{a}{b}$ where a and b are integers and $b \neq 0$. Equality of rational numbers is defined by $\dfrac{a}{b} = \dfrac{c}{d}$ if and only if $ad = bc$.

Addition and multiplication of rational numbers are defined in terms of addition and multiplication of integers by the laws:

$$\frac{a}{b} + \frac{c}{d} = \frac{(a \times d) + (b \times c)}{b \times d}$$

$$\frac{a}{b} \times \frac{c}{d} = \frac{a \times c}{b \times d}$$

All of the field postulates can easily be checked. We shall check the commutative property of addition. We compute

$$\frac{a}{b} + \frac{c}{d} = \frac{(a \times d) + (b \times c)}{b \times d}$$

and

$$\frac{c}{d} + \frac{a}{b} = \frac{(c \times b) + (d \times a)}{d \times b}$$

These two expressions are equal since the commutative property holds for addition and multiplication of integers.* To check that every rational number except zero has a multiplicative inverse, we note that the multiplicative inverse (**recip-rocal**) of $\dfrac{a}{b}$ is the rational number $\dfrac{b}{a}$ (for every number except $\dfrac{0}{b}$, $b \neq 0$).

Example 2. Let a system be defined to consist of a set S whose elements are two distinct words, "even" and "odd." We define

$$
\begin{array}{ll}
\text{odd} + \text{odd} = \text{even}; & \text{odd} \times \text{odd} = \text{odd} \\
\text{odd} + \text{even} = \text{odd}; & \text{odd} \times \text{even} = \text{even} \\
\text{even} + \text{odd} = \text{odd}; & \text{even} \times \text{odd} = \text{even} \\
\text{even} + \text{even} = \text{even}; & \text{even} \times \text{even} = \text{even}
\end{array}
$$

We can more easily verify that this system is a field if we make an operation table (Table 6.1) showing the operations just defined. It is easy to verify that this system is a field. The

* See Chapter 2, Section 1.

TABLE 6.1. *Operation table for even-odd*

+	Even	Odd		×	Even	Odd
even	even	odd		even	even	even
odd	odd	even		odd	even	odd

system is certainly closed under both operations. Both operations are commutative and associative. The identity for addition is even, and the identity for multiplication is odd. Each element is its own additive inverse. The multiplicative inverse of odd is odd. This field is a small one and does not cause much difficulty in checking all the postulates.

EXERCISE 2

1. Verify that multiplication of rational numbers is commutative.
2. Verify that addition of rational numbers is associative.
3. Verify that multiplication of rational numbers is associative.
4. Verify the distributive property for the set of rational numbers.
5. Verify that $\dfrac{0}{b}$, $b \neq 0$, is the additive identity for the set of rational numbers.
6. Verify that $\dfrac{b}{b}$, $b \neq 0$, is the multiplicative identity for the set of rational numbers.
7. Using Table 6.1 verify that:
   (a) even + (odd + odd) = (even + odd) + odd
   (b) even × (odd + odd) = (even × odd) + (even × odd)
8. Verify that even is the additive identity for the even-odd system.
9. Verify that odd is the multiplicative identity for the even-odd system.

3. THEOREMS FOR A FIELD

Various trivial theorems can be proved at once from the field postulates. Most of us know these theorems. We learned them when we studied arithmetic and elementary algebra.

THEOREM 6.1: If a is an element of F, then $a \times 0 = 0$.

Proof: Assume that b is an element of F. Then

$$a \times (b + 0) = (a \times b) + (a \times 0)$$
Distributive Property

$$a \times b = (a \times b) + (a \times 0)$$
Additive Identity

Since every element in F has an additive inverse, $-(a \times b)$ exists, and because of closure for addition we have

$$-(a \times b) + (a \times b) = -(a \times b) + [(a \times b) + (a \times 0)]$$
$$0 = [-(a \times b) + (a \times b)] + (a \times 0)$$
Additive Inverses and Associative Property of Addition

$$0 = 0 + (a \times 0) \quad \text{Additive Inverses}$$
$$0 = a \times 0 \quad\qquad \text{Additive Identity}$$
$$a \times 0 = 0 \quad\qquad \text{Symmetry for equality}$$

THEOREM 6.2: $(-1) \times a = -a$.

Proof: $a + (-a) = 0$ Additive Inverses
$$= a \times 0 \qquad\qquad \text{Theorem 6.1}$$
$$= [1 + (-1)] \times a \quad \text{Additive Inverse and Commutative under Multiplication}$$
$$= (1 \times a) + [(-1) \times a] \quad \text{Distributive Property}$$

As a consequence of the foregoing steps we have

$$(-a) + [a + (-a)] = (-a) + \{(1 \times a) + [(-1) \times a]\}$$
$$[(-a) + a] + (-a) = [(-a) + a] + [(-1) \times a]$$
$$0 + (-a) = 0 + [(-1) \times a]$$
$$-a = (-1) \times a$$
$$(-1) \times a = -a$$

The foregoing steps may be verified by the closure property of addition and additive inverses, associative property of addition and multiplicative identity, additive inverses, the additive identity, and the symmetric property of equality.

THEOREM 6.3: If $a \times b = 0$, then $a = 0$ or $b = 0$.

Proof: $a = 0$ or $a \neq 0$. If $a = 0$, the theorem is true. If $a \neq 0$, it possesses a multiplicative inverse a^{-1} in F by the multiplicative inverse postulate. Hence

$$a^{-1} \times (a \times b) = a^{-1} \times 0$$
$$(a^{-1} \times a) \times b = a^{-1} \times 0$$
$$1 \times b = a^{-1} \times 0$$
$$b = a^{-1} \times 0$$
$$b = 0$$

These steps may be verified by closure for multiplication, associative property of multiplication, multiplicative inverses, multiplicative identity, and Theorem 6.1. The foregoing shows that if $a \neq 0$, then $b = 0$; that is, either a or b must be 0.

THEOREM 6.4: $-(-x) = x$.

Proof:

$$-(-x) + (-x) = 0$$
$$x + (-x) = 0$$
$$-(-x) + (-x) = x + (-x)$$
$$[-(-x) + (-x)] + x = [x + (-x)] + x$$
$$-(-x) + [(-x) + x] = [x + (-x)] + x$$
$$-(-x) + 0 = 0 + x$$
$$-(-x) = x$$

The reader should verify each step with a postulate, theorem, or property of the equivalence relation equality.

EXERCISE 3

1. In the set of rational numbers what is the multiplicative inverse of each of the following?

   (a) $\dfrac{3}{5}$ (d) $\dfrac{-6}{7}$

   (b) $\dfrac{3}{4}$ (e) $\dfrac{-7}{15}$

   (c) $\dfrac{7}{-6}$ (f) $\dfrac{19}{12}$

2. In the set of rational numbers what is the additive inverse of each of the following?

   (a) $\dfrac{1}{2}$ (d) $-\dfrac{2}{3}$

   (b) $\dfrac{3}{4}$ (e) $-\dfrac{1}{2}$

   (c) $-\dfrac{1}{4}$ (f) $\dfrac{7}{9}$

3. In the accompanying table check the eleven postulates for a field for each of the given sets. Place a "+" in the chart if the postulate is satisfied, and a "0" if the postulate is not satisfied.

	1	2	3	4	5	6	7	8	9	10	11
(a) All positive rational numbers.											
(b) All even positive integers.											
(c) All odd positive integers.											
(d) All negative integers.											
(e) All real numbers.											
(f) $\{0, 1, 2, -2, -1\}$.											
(g) All integral multiples of 5.											
(h) All integral multiples of 3.											

4. If t, y, and z are elements of a field, prove that if $t \times y = t \times z$ and $t \neq 0$, then $y = z$.

5. If t, y, and z are elements of a field, prove that if $t + y = t + z$, then $y = z$.

6. If t and y are elements of a field, prove $t \times (-y) = -(t \times y)$.

7. Prove that there is only one additive identity in a field. (*Hint:* Assume that there are two and prove that they are equal.)

8. Prove that there is only one multiplicative identity in a field. (*Hint:* Assume there are two and prove that they are equal.)

9. Show that a single element y subject to the laws $y + y = y$ and $y \times y = y$, forms a field.

10. Prove that if x and y are elements of a field $(-x) \times (-y) = x \times y$.

4. GROUPS

A less familiar abstract mathematical system than a field is a group. A **group**, G, is a set of elements, $\{a, b, c, \ldots\}$, with one binary operation denoted by the symbol "$*$" satisfying the following postulates:

G-1. (*Closure*): If a and b are elements of G, then $a * b$ is an element of G.

G-2. (*Associative*): If a, b, and c are elements of G, then

$$(a * b) * c = a * (b * c)$$

G-3. (*Identity Element*): There exists an element e of G, called the **identity element** such that for all elements a of G

$$a * e = e * a = a$$

G-4. (*Inverse Elements*): For every element a of G there is an element a^{-1} called the **inverse** of a such that

$$a * a^{-1} = a^{-1} * a = e$$

where e is the identity element.

If, in addition to the foregoing postulates, a group satisfies the following postulate, then we say that the group is a **commutative** or **Abelian group**.

G-5. (*Commutative*): If a and b are elements of G, then

$$a * b = b * a$$

Recall that the set of integers does not satisfy all the postulates of a field under the operations of addition and multiplication. This set, how-

ever, does satisfy all four postulates for a group with the operation *
interpreted as addition. We know that if we add two integers the result is
an integer, so G-1 is satisfied.

Postulate G-2 demands that the associative property of addition
holds for the set of integers. From previous experience we know this is true.

G-3 is satisfied by using the number 0 for e.

Postulate G-4 is satisfied. Given an integer, its negative is its inverse
with respect to addition.

Since all the group postulates are satisfied with the operation of
addition, we say that the set of integers is a group under addition or the
set of integers is a group with respect to addition.

Since addition is commutative for the set of integers, we say that
the set of integers with the operation of addition is a commutative or
Abelian group.

Some other examples of groups are given below.

Example 1. The set of all rational numbers different from zero and the
operation of multiplication. Zero must be omitted since it
has no multiplicative inverse.

Example 2. The set $\{0, 1, 2, 3, 4\}$ and addition defined modulo 5.*

Postulate G-1 and Postulate G-2 are certainly satisfied.
Postulate G-3 is satisfied taking $e = 0$. Postulate G-4 is
satisfied since

$$2 + 3 \equiv 0 \,(\text{mod } 5)$$
$$1 + 4 \equiv 0 \,(\text{mod } 5)$$
$$0 + 0 \equiv 0 \,(\text{mod } 5)$$

Since addition is commutative for modular systems, this is a
commutative group.

EXERCISE 4

1. Why is the set of positive integers not a group with respect to
   addition?

* See Chapter 4, Section 1.

2. Why is the set of integers not a group with respect to multiplication?

3. Which of the following sets are groups with respect to the given operations?

Set	Operation
(a) nonnegative integers	addition
(b) positive integers	multiplication
(c) odd integers	addition
(d) whole numbers	multiplication
(e) 0, 1, 2, 3	addition (mod 4)
(f) rational numbers	addition
(g) 1, 2, 3, 4, 5, 6,	multiplication (mod 7)
(h) 0, 1, 2, 3	multiplication (mod 4)

4. Let set $S = \{1, 2, 3, 4\}$ and the binary operation $*$ defined in the table below.

*	1	2	3	4
1	1	2	3	4
2	2	4	1	3
3	3	1	4	2
4	4	3	2	1

(a) Is this system a group? Why or why not?

(b) What is the inverse of each element in S?

5. Explain why each of the following operation tables do not exhibit a group. Give as many reasons as possible.

(a)

*	x	y	z
x	x	y	z
y	y	x	z
z	x	x	x

(b)

*	a	b
a	b	a
b	b	b

6. Prove that the set of all integral multiples

$$\{\cdots -2m, -m, 0, m, 2m, \ldots\}$$

of a fixed integer m is a commutative group with respect to the operation of addition.

7. Show that the numbers 1, 3, 5, and 7, with multiplication defined modulo 8, form a group.

8. Prove that the identity element of a group is unique. (*Hint*: Assume

that there are two and prove that they are equal.)
9. Prove that the inverse element of any element a of a group G is unique.

5. GROUPS WITHOUT NUMBERS

We shall exhibit an example of what is called a **group of trans-formations**. One definition of geometry is: geometry is the study of properties of figures that are unchanged by applying transformations from a group of transformations. In Euclidean geometry properties of length, area, volume, and measurement of angles are not changed when the figures are moved about freely in space (assuming that the figures are rigid). These movements can be described by forming a group of trans-formations.

The particular group of transformations we shall consider has elements that are movements of a square. There are eight movements involved. We shall call these movements $R_0, R_1, R_2, R_3, H, V, D,$ and D'. Think of a square with corners numbered as shown in Figure 6.1. This square region has the same numerals on the back of it directly under the numerals displayed.

figure 6.1

We define R_0 to be the movement that rotates the square 360 degrees in a clockwise direction. Thus

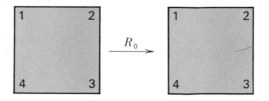

R_1 means to rotate the square 90 degrees in a clockwise direction. Thus

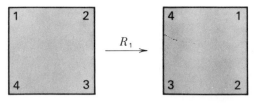

R_2 means to rotate the square 180 degrees in a clockwise direction. Thus

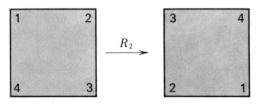

R_3 means to rotate the square 270 degrees in a clockwise direction. Thus

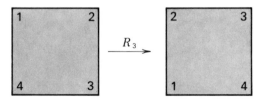

H means to rotate the square around its horizontal axis. Thus

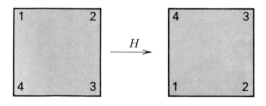

V means to rotate the square around its vertical axis. Thus

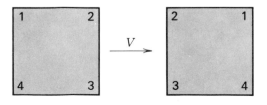

D means to rotate the square about the diagonal from the bottom left corner to the top right corner. Thus corners numbered 2 and 4 remain fixed and corners numbered 1 and 3 change places:

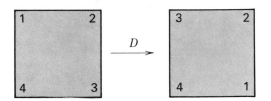

D' means to rotate the square about the diagonal from the top left corner to the bottom right corner. Thus

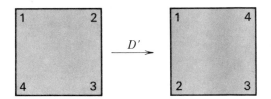

We define the operation * as follows: $H * V$ means to perform the transformation (or movement) H, and then on the result perform the transformation V. Thus

$$H * V = R_2$$

$R_1 * R_2$ means to perform R_1, and then on the result perform R_2. Thus

$$R_1 * R_2 = R_3$$

We must be careful of the order here. $H * V$ does not mean the same as $V * H$. $H * V$ means do H first, V second. $V * H$ means do V first, H second. Thus the symbol $V * H$ may be read: transformation V followed by transformation H. Care is needed because the end results may not be the same.

An operation table now may be made to exhibit all the results. This operation table is shown in Table 6.2. The elements in the left column of the table represent the first movement or transformation performed; those in the top row, the second movement or transformation. (A good way to perform these operations is to use a cardboard square region and do the actual movements with it. Be sure that the numerals in the corners are on both sides.)

TABLE 6.2. *Operation table for the transformation of a square*

	*	R_0	R_1	R_2	R_3	H	V	D	D'
					Second Transformation				
First Transformation	R_0	R_0	R_1	R_2	R_3	H	V	D	D'
	R_1	R_1	R_2	R_3	R_0	D	D'	V	H
	R_2	R_2	R_3	R_0	R_1	V	H	D'	D
	R_3	R_3	R_0	R_1	R_2	D'	D	H	V
	H	H	D'	V	D	R_0	R_2	R_3	R_1
	V	V	D	H	D'	R_2	R_0	R_1	R_3
	D	D	H	D'	V	R_1	R_3	R_0	R_2
	D'	D'	V	D	H	R_3	R_2	R_1	R_0

We must now verify that this system is a group. Table 6.2 exhibits the property of closure: no results are obtained except the eight elements of the set of transformations of a square. It also exhibits the presence of an element R_0 which has no effect on the other elements. Hence R_0 is the identity. That the associative property holds can be verified by examining all the possible cases.* To illustrate this verification let us examine a few cases.

(1) $(R_1 * R_2) * H = R_3 * H = D'$
    $R_1 * (R_2 * H) = R_1 * V = D'$

$$\therefore (R_1 * R_2) * H = R_1 * (R_2 + H)$$

(2) $(V * D) * D' = R_1 * D' = H$
    $V * (D * D') = V * R_2 = H$

$$\therefore (V * D) * D' = V * (D * D')$$

(3) $(R_3 * V) * H = D * H = R_1$
    $R_3 * (V * H) = R_3 * R_2 = R_1$

$$\therefore (R_3 * V) * H = R_3 * (V * H)$$

We must now show that every element has an inverse. This also can be found from the table. The inverse of H is H because $H * H = R_0$ (the identity element); the inverse of V is V since $V * V = R_0$; the inverse of

* There are 512 cases required to prove the associative property.

D is D because $D * D = R_0$; the inverse of D' is D' because $D' * D' = R_0$; the inverse of R_1 is R_3 because $R_1 * R_3 = R_0$; the inverse of R_2 is R_2 because $R_2 * R_2 = R_0$; the inverse of R_3 is R_1 because $R_3 * R_1 = R_0$.

This group of transformations of a square is not a commutative group as can be seen since, in particular,

$$H * D \neq D * H$$

EXERCISE 5

1. From Table 6.2 compute:
   (a) $R_1 * (H * R_3)$ (d) $(H * D') * R_2$
   (b) $(V * R_1) * R_3$ (e) $(H * V) * (D * R_3)$
   (c) $(D * D') * D$ (f) $(H * V) * (D * D')$

2. Make a mathematical system for the transformations of an isosceles triangle. There will be two movements: (1) a turn of 360 degrees in a clockwise direction (call it I), and (2) a turn around the vertical axis (call it V). Make an operation table of this system. Is it a group? If so, is it a commutative group?

3. Construct a mathematical system for the movements of an equilateral triangle. Make a triangle and label the vertices as shown in the diagram. There will be six transformations. There will be three rotations in a clockwise direction: one of 120 degrees, one of 240 degrees, and one of 360 degrees. There will be three rotations about the axes (broken lines in the figure). Make an operation table for this system. Is it a group? If so, is it a commutative group?

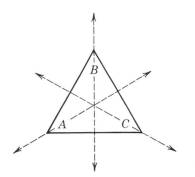

4. Construct a mathematical system for the transformations of a regular pentagon. Make a pentagon and label it as shown in the diagram on the next page.

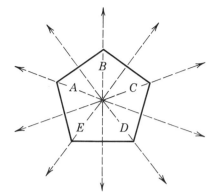

There will be ten transformations. There will be five rotations about the axes shown by broken lines in the figure, and five rotations in a clockwise direction of 72, 144, 216, 288, and 360 degrees. Make an operation table for this system. Is it a group?

6. SOME THEOREMS ABOUT GROUPS

We will now prove some theorems about groups.

THEOREM 6.5: If a, b, and c are elements of a group G, and $c = b$, then $a * b = a * c$.

Proof: We know that

$$a * c = a * c$$

by the reflexive law of equality. Since

$$c = b$$

we know

$$a * b = a * c$$

by the law of substitution.*

THEOREM 6.6: The identity element of a group G is unique.

Proof: By G-3 we know that there is an identity element in any group. Call this identity element e_1. We wish to show that e_1 is the only identity element in the group. Let us assume that there is another identity element e_2.

* See Chapter 1, Section 11.

Since e_1 is an identity element

$$e_1 * e_2 = e_2$$

Since e_2 is an identity element

$$e_1 * e_2 = e_1$$

Hence we have

$$e_1 = e_1 * e_2 \qquad \text{and} \qquad e_1 * e_2 = e_2$$

and by the transitive property of equality we have

$$e_1 = e_2$$

THEOREM 6.7: **The inverse element of any element of a group G is unique.**

Proof: By G-4 we know that for any element a of G there is an inverse a^{-1}. Suppose that there are two inverse elements for a. Call them a_1^{-1} and a_2^{-1}. Then

$$a * a_1^{-1} = e$$

and

$$a * a_2^{-1} = e$$

Hence

$$a * a_1^{-1} = a * a_2^{-1}$$
$$u_1^{-1} * (a * a_1^{-1}) = a_1^{-1} * (a * a_2^{-1})$$
$$(a_1^{-1} * a) * a_1^{-1} = (a_1^{-1} * a) * a_2^{-1}$$
$$e * a_1^{-1} = e * a_2^{-1}$$
$$a_1^{-1} = a_2^{-1}$$

THEOREM 6.8: **If a and b are elements of a group G, and $a * b = a * c$, then $b = c$.**

Proof: We have given that $a * b = a * c$, hence by G-1 and G-4 we have

$$a^{-1} * (a * b) = a^{-1} * (a * c)$$
$$(a^{-1} * a) * b = (a^{-1} * a) * c \quad \text{(by G-2)}$$

Hence by G-4

$$e * b = e * c$$

and

$$b = c$$

by G-3.

EXERCISE 6

Prove the following theorems:

1. If x and y are elements of a group G, then $(x * y) * y^{-1} = x$.
2. If a and b are elements of a group G, and $a * x = a$, then $x = e$.
3. In a group G, $e = e^{-1}$.
4. If a and b are elements of a group G, then $(a * b)^{-1} = b^{-1} * a^{-1}$.
5. If a and b are elements of a group G and $a = b$, then $a^{-1} = b^{-1}$.
6. In a group G, $a * x = b$ has a unique solution $x = a^{-1} * b$. (You must prove two things: (1) if $a * x = b$, then $x = a^{-1} * b$; and (2) show that $x = a^{-1} * b$ is indeed a solution of $a * x = b$.)

7. PERMUTATION GROUPS

Let us consider the three letters A, B, C. How many different arrangements can we make of these three letters? We could, of course, write out all the arrangements of the letters A, B, C. If we did, we would find that there were six arrangements.

Now let us solve this problem by a method that will provide a rule for finding how many different arrangements we can make from n letters.

Let us suppose that we have three boxes into each of which we put one of the letters A, B, C.

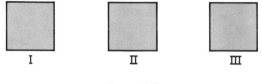

I   II   III

figure 6.2

In Box I we can put any one of the three letters. Hence the choice of the first letter can be made in three ways. To show this we write "3" in Box I.

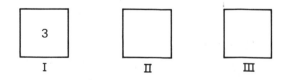

I   II   III

figure 6.3

Having put one of the letters in Box I, we have two letters left. We can put any one of these two in Box II; that is, we can fill Box II in two ways. To show this we write "2" in Box II. This means that altogether there are 3 × 2 ways of filling Boxes I and II. Since only one element is left, there is only one way to fill Box III. To show this we write "1" in Box III.

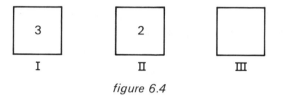

figure 6.4

We now have 3 × 2 × 1 = 6 arrangements that can be made from the letters *A*, *B*, and *C*.

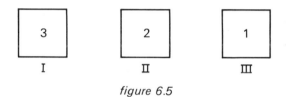

figure 6.5

We call each arrangement of the three letters a **permutation**. There are six permutations of three elements.

Now let us generalize the foregoing example and find how many permutations we can make of *n* elements. Suppose we have *n* elements. Let us imagine that we have *n* boxes into each of which we can put one element. In Box I we can put any one of *n* elements. Having put one of the elements in Box I, we have $(n - 1)$ elements left; so we can put any one of these in Box II. That is, we can fill Box II in $(n - 1)$ ways. This means that together there are $n(n - 1)$ ways to fill Boxes I and II. We now have $(n - 2)$ elements left; so we can fill Box III in $(n - 2)$ ways. This means that together there are $n(n - 1)(n - 2)$ ways of filling Boxes I, II, and III. Continuing in this fashion we find that the number of permutations of the *n* elements is

$$n(n - 1)(n - 2)(n - 3) \ldots (2)(1)$$

In general, then, the number of permutations of *n* elements, *n* a natural number, taken *n* at a time is

$$n(n - 1)(n - 2)(n - 3) \ldots (2)(1)$$

We use the symbol $n!$ to denote the product $n(n - 1)(n - 2)\ldots$ $(2)(1)$. We read this symbol "**n factorial**." Thus we see that

$$5! = 5 \times 4 \times 3 \times 2 \times 1 = 120$$
$$4! = 4 \times 3 \times 2 \times 1 = 24$$
$$3! = 3 \times 2 \times 1 = 6$$

We define $1! = 1$. From the preceding we see that the number of permutations of n elements taken n at a time is $n!$. It should be emphasized that *in a permutation we are concerned with the order of the elements*. The permutation ABC is not the same as BCA.

Let us call the six permutation of three elements e, p, q, r, s, and t as follows:

$$e = \begin{pmatrix} A & B & C \\ A & B & C \end{pmatrix} \quad p = \begin{pmatrix} A & B & C \\ C & A & B \end{pmatrix} \quad q = \begin{pmatrix} A & B & C \\ B & C & A \end{pmatrix}$$

$$r = \begin{pmatrix} A & B & C \\ B & A & C \end{pmatrix} \quad s = \begin{pmatrix} A & B & C \\ C & B & A \end{pmatrix} \quad t = \begin{pmatrix} A & B & C \\ A & C & B \end{pmatrix}$$

These form the set G. Now we define an operation $*$ on these elements. Let $p * q$ mean that we first carry out p and then q. We see that

$$p * q = \begin{pmatrix} A & B & C \\ C & A & B \end{pmatrix} * \begin{pmatrix} A & B & C \\ B & C & A \end{pmatrix} = \begin{pmatrix} A & B & C \\ A & B & C \end{pmatrix} = e$$

for p replaces A by C, and q replaces C by A, hence A is finally replaced by A; p replaced B by A, and q replaces A by B, hence B is finally replaced by B; p replaces C by B, and q replaces B by C, hence C is finally replaced by C. But this arrangement is e. Hence $p * q = e$.

Using the same definition of the operation $*$ we have

$$r * p = \begin{pmatrix} A & B & C \\ B & A & C \end{pmatrix} * \begin{pmatrix} A & B & C \\ C & A & B \end{pmatrix} = \begin{pmatrix} A & B & C \\ A & C & B \end{pmatrix} = t$$

$$p * r = \begin{pmatrix} A & B & C \\ C & A & B \end{pmatrix} * \begin{pmatrix} A & B & C \\ B & A & C \end{pmatrix} = \begin{pmatrix} A & B & C \\ C & B & A \end{pmatrix} = s$$

With this understanding G is a group under the operation $*$. It is called a **permutation group**. It is not a commutative group for, in particular, $r * s \neq s * r$.

EXERCISE 7

1. Complete the following operation table for the permutation group just discussed.

		Second assignment					
	*	e	p	q	r	s	t
First Assignment	e	e	p	q	r	s	t
	p	p		e	s		
	q	q					
	r		t			q	
	s	s					
	t	t					

2. Using the table constructed in problem 1, compute:
   (a) $(r * s) * t$ (d) $(s * t) * (p * q)$
   (b) $(q * p) * e$ (e) $(s * t) * (p * r)$
   (c) $(r * q) * q$ (f) $(t * q) * (p * r)$
3. What is the identity element for the above permutation group?
4. Give the inverse element for every element in the foregoing permutation group
5. How many elements would be in a permutation group when:
   (a) Two elements are permuted?
   (b) Five elements are permuted?
   (c) Four elements are permuted?
6. Construct an operation table for the permutation of two elements.

$$e = \begin{pmatrix} A & B \\ A & B \end{pmatrix} \qquad p = \begin{pmatrix} A & B \\ B & A \end{pmatrix}$$

7. Form all the permutations from a set of four elements, A, B, C, and D.
8. Form the operation table for the twenty-four permutations of four elements.
9. Compute the following.

$$\begin{pmatrix} A & B & C & D \\ D & C & B & A \end{pmatrix} * \begin{pmatrix} A & B & C & D \\ A & C & D & B \end{pmatrix}$$

10. What is the inverse of each element in the permutation group in problem 8?

11. Let *ABC* be an equilateral triangle with vertices named clock-wise, and let transformations of this triangle into itself be described by the following:

(i) 120° clockwise rotation about the center;
(ii) 240° clockwise rotation about the center;
(iii) 360° clockwise rotation about the center;
(iv) 180° flipping about the line of symmetry passing through *A*;
(v) 180° flipping about the line of symmetry passing through *B*;
(vi) 180° flipping about the line of symmetry passing through *C*.

(a) Use the two-row notation of Section 7 to describe replace-ments of the vertices in each transformation. For example, (iv) is illustrated below:

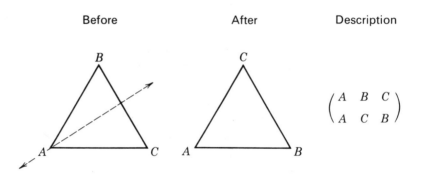

Before   After   Description

$$\begin{pmatrix} A & B & C \\ A & C & B \end{pmatrix}$$

(b) Compare these descriptions with the six permutations in Section 7 and label these transformations accordingly.

JEROME CARDAN (1501–1576) *or Girolamo Cardano, the illegitimate son of an Italian jurist, was a pioneer in the field of probability. At one time or another he was a professor at the University of Bologna, the rector of the College of Physicians of Milan, and an inmate of an almshouse. He was indeed a man of contrasts. He was an astrologer as well as a student of philosophy, a gambler as well as an algebraist, a physician as well as the father of a murderer, and a man of genius as well as a man devoid of principles. Cardan's book,* Liber de Ludo Aleae (Book on Games of Chance), *is really a gambler's manual and may be considered the first book on probability. In this book Cardan discussed how many successes in how many trials, mathematical expectation, and the additive properties of probability. He correctly showed the number of ways that two dice may be thrown and also listed the ways in which three dice may be thrown. As a dramatic ending to this unprincipled genius, Cardan predicted the day of his own death. When the day arrived and he was still alive, he committed suicide to make his prediction come true.*

Probability
and Statistics/7

1. PROBABILITY MEASURE

Suppose we roll a die (plural: dice). Let us assume that the die is perfectly balanced, and it is as likely to fall with one face showing as another. Such a die is called an "honest" die. There are six possible **outcomes**: 1, 2, 3, 4, 5, or 6 showing. The set S of all outcomes is called a **sample space** of the experiment.

$$S = \{1, 2, 3, 4, 5, 6\}$$

Now let us consider some subsets of this sample space. For example, we might consider the subset E_1 of even numbers showing:

$$E_1 = \{2, 4, 6\}$$

Or we might consider the subset E_2 of numbers greater than 4 showing:

$$E_2 = \{5, 6\}$$

Such subsets, when defined by a special condition, are called **events** in the sample space.

For example, the event E, "an odd number showing," is

$$E = \{1, 3, 5\}$$

The **probability** of an event is a number between 0 and 1 inclusive. If an event is certain to happen, its probability is 1. If an event cannot happen its probability is zero. In a situation with several possible outcomes, all equally likely, *the probability of an event is equal to the number* *of elements in the event divided by the number of elements in the sample space.*

Thus the probability of the event F, "a 6 showing," is $\frac{1}{6}$. The probability of the event E_1, "an even number showing," is $\frac{3}{6} = \frac{1}{2}$.

Instead of writing the word "probability" each time, we represent it by the letter P. For example, for the probability of the event E we write $P(E)$. Thus for the probability of the event E_1, "an even number showing," we write

$$P(E_1) = \tfrac{3}{6} = \tfrac{1}{2}$$

If E is an event in a sample space, the notation E' may be used to state that the event E does not occur. In agreement with the law of the excluded middle,* in any situation either E or E' occurs and

$$P(E) + P(E') = 1$$

For example, if E is the event, "a 4 is showing," then $P(E) = \frac{1}{6}$. The probability of the event E', "a 4 is not showing," is $P(E') = \frac{5}{6}$. And

$$P(E) + P(E') = \tfrac{1}{6} + \tfrac{5}{6} = 1$$

EXERCISE 1

1. Suppose you toss a coin. What is a sample space?
2. If you toss a coin, what is the probability of the event:
   (a) heads   (c) not heads
   (b) tails   (d) not tails
3. If you toss a die, what is the probability of the event:
   (a) a 3 showing
   (b) a 6 showing
   (c) a number greater than 3 showing
   (d) a number less than 5 showing
   (e) an odd number showing
4. If a die has been rolled twenty-five times and has not shown a 5, what is the probability that it will show a 5 on the next roll?
5. Suppose you draw a card from a deck of bridge cards. How many outcomes are in a sample space?
6. Suppose you draw a card from a deck of bridge cards. What is the probability that you will draw a king of hearts?
7. Suppose you draw a card from a deck of bridge cards. What is the probability that you will draw an ace?
8. Suppose you draw a card from a deck of bridge cards? What is the probability that you will draw:

* See Chapter 1, Section 9.

(a) a spade

(b) a face card (jack, queen, or king)

(c) a deuce.

9. A poll taker reports that the probability that candidate X will win an election is $\frac{3}{5}$, while the probability of his losing is $\frac{1}{4}$. Is this possible? Why or why not?

10. If the probability that a given stock will increase in price is $\frac{1}{4}$, what is the probability that it will not rise in price?

11. A poll is taken among fifty employees at an industrial plant on the question of having a union shop. The results are shown in the accompanying table.

	Supervisors	Hourly Employees	Salaried Employees	Totals
For	1	30	1	32
Against	4	3	6	13
No opinion	0	2	3	5
Totals	5	35	10	50

(a) What is the probability that an employee selected at random will be a supervisor? (An item is selected at **random** from a group of items if the selection procedure is such that each item in the group is equally likely to be selected.)

(b) What is the probability that an employee selected at random will be against union affiliation?

(c) What is the probability that an employee selected at random will be an hourly worker who favors the union?

12. At a gambling casino there is a wheel with the numerals 1, 2, 5, 10, and 20 marked on it. A wheel similar to the wheel in the casino is shown in the accompanying figure.

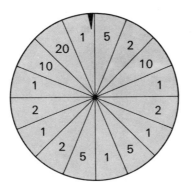

What is the probability of the following events?

(a) 1 will appear under the pointer.

(b) 2 will appear under the pointer.

(c) 5 will appear under the pointer.

(d) 10 will appear under the pointer.

(e) 20 will appear under the pointer.

13. Which of the following statements are true?

(a) Events with the same number of elements have equal probabilities.

(b) The smaller the probability, the less likely it is that an event will occur.

(c) The larger the probability the less likely it is that an event will occur.

(d) The probability of the event "the number 7 showing" when a die is tossed is 0.

2. EXPERIMENT WITH A PAIR OF DICE

Suppose we have two dice, one red and one green. Now let us perform an experiment of rolling these two dice. What are the possible outcomes?

The outcomes on the red (R) die are 1, 2, 3, 4, 5, and 6. The outcomes on the green (G) die are 1, 2, 3, 4, 5, and 6. The method for finding all the possible outcomes in our experiment of rolling two dice is shown in Figure 7.1. Such a diagram is called a **tree**. The lines in the diagram represent the possible outcomes. For instance, 1 on the red (R) die may be matched with six possibilities on the green (G) die. We see from the figure that the set of all possible outcomes—that is, a sample space— has $6 \times 6 = 36$ elements.

We shall designate these outcomes as (11), (12), (13), (14), and so forth. Thus (45) means 4 on the red die and 5 on the green die.

Now we ask what is the probability of each outcome in the sample space? Since there are thirty-six possible outcomes, and each is equally likely to occur, the probability of each outcome is $\frac{1}{36}$. We may write

$$P(11) = \frac{1}{36}$$
$$P(34) = \frac{1}{36}$$
$$P(56) = \frac{1}{36}$$
$$\vdots$$

$$R \qquad\qquad G$$

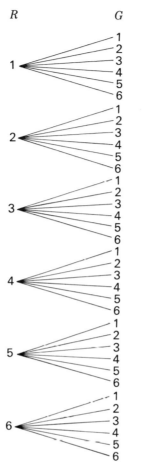

figure 7.1

Now let us look at the sums that will appear on the dice when they are rolled. The possible sums are 2, 3, 4, 5, 6, 7, 8, 9, 10, 11, and 12. Let us see how many times each sum can occur (Table 7.1).

There are many events we might consider concerning this experiment of rolling a pair of dice. Let us consider the event E, "obtaining a sum of 7." Then

$$E = \{(16), (25), (34), (43), (52), (61)\}$$

The number of elements in E is 6. The number of elements in the sample space is 36. Hence

$$P(E) = \tfrac{6}{36} = \tfrac{1}{6}$$

TABLE 7.1

Outcome	Sum	Number of Appearances in the Sample Space
(11)	2	1
(12), (21)	3	2
(13), (22), (31)	4	3
(14), (23), (32), (41)	5	4
(15), (24), (33), (42), (51)	6	5
(16), (25), (34), (43), (52), (61)	7	6
(26), (53), (44), (53), (62)	8	5
(36), (45), (54), (63)	9	4
(46), (55), (64)	10	3
(56), (65)	11	2
(66)	12	1

Since the number of elements in the event F, "obtaining a sum of 11," is two,

$$F = \{(56), (65)\}$$

we see that

$$P(F) = \tfrac{2}{36} = \tfrac{1}{18}$$

EXERCISE 2

1. Find the probability of the following events for the experiment of rolling two dice.
   (a) Obtaining a sum of 12.
   (b) Obtaining a sum of 6.
   (c) Obtaining a sum of 10.
   (d) Obtaining a sum of 8.
   (e) Obtaining a sum that is an even number.
   (f) Obtaining a sum greater than 5.
2. A penny, a nickel, a dime, and a quarter are tossed at once.
   (a) List all the possible outcomes. (*Hint:* Use a tree diagram, a portion of which appears on the next page.)

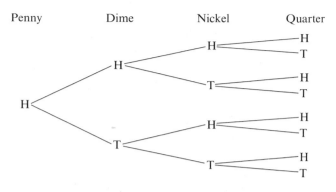

Penny   Dime   Nickel   Quarter

(b) What is the probability of four heads?
(c) What is the probability of three heads and a tail?
(d) What is the probability of four tails?
(e) What is the probability of two heads and two tails?
3. A coin and a die are tossed at the same time.
   (a) List all the possible outcomes.
   (b) What is the probability of the coin falling heads and the die showing 6?
   (c) What is the probability of the coin falling heads?
   (d) What is the probability of the coin falling heads and the die showing a number greater than 4?
4. Give the probabilities of the six possible outcomes when five coins are tossed all together. (*Hint:* HHHTT and HTTHH are the same outcomes: three heads and two tails.)
5. You pick a 9 and a 7 from a deck of bridge cards. You return them to the deck. What is the probability that you will draw a 5 on the third draw?
6. There are twenty slips of paper in a hat. Each slip has a different name written on it. A slip of paper is drawn at random from the hat, the name on it read, and the slip returned to the hat. Let the names on the slips be represented by the first twenty letters of the alphabet: $A, B, C, \ldots, S, T$.
   (a) Is $P(A) = P(R)$?
   (b) What is $P(S) + P(J)$?
   (c) What is $P(H) + P(Q) + P(B)$?
   (d) What is $P(A) + P(B) + P(C) + \cdots + P(T)$?
7. Two regular tetrahedra (a regular tetrahedron is a regular solid having four congruent faces) have faces marked with 1, 2, 3, and 4. They are tossed in the air and allowed to fall on a table. We describe an outcome by the numeral on the bottom face.

(a) How many possible outcomes are there?

(b) What is the probability of the event "a sum of 8"?

(c) What is the probability of the event "a sum of 5"?

(d) How many possible outcomes are there if three such tetrahedra are tossed? What is the probability of the event "two 3's and one 5"?

8. A box contains ten beads of which five are black, three are white, and two are blue. All the beads are identical in size, hence each bead is equally likely to be picked if you reach in the box and take one bead at random.

(a) What is the probability that you will pick a black bead?

(b) Assuming you pick a black bead and return it to the box, what is the probability that you will pick a black bead on the second draw?

(c) Assume that you pick a black bead on the first draw and a white bead on the second draw. You return both beads to the box. What is the probability that you will pick a blue bead on the third draw?

9. Two regular octahedra (a regular octahedron is a regular solid having eight congruent faces) have faces numbered serially 1 through 8. We toss them and allow them to fall on a hard surface. We describe an outcome by the numerals on the bottom faces. What is the probability of the event:

(a) a sum of 8

(b) a sum of 16

(c) a sum of 7

(d) a sum of 11.

10. Two pairs of shoes have been thrown on the floor of a dark closet. You pick up the first two shoes you feel. What is the probability that you will have a matched pair?

3. MUTUALLY EXCLUSIVE EVENTS

In computing probabilities, the relation of events to each other must be taken into consideration. If S is a sample space and E_1 and E_2 are events relative to S, we can have the following cases:

1. E_1 and E_2 have no elements in common.
2. E_1 and E_2 have some common elements.

The set of elements common to E_1 and E_2 is denoted by $E_1 \cap E_2$, read: "the intersection of E_1 and E_2" or simply "the event E_1 and E_2."

If E_1 and E_2 have no elements in common we say that E_1 and E_2 are **mutually exclusive events**. If E_1 and E_2 are mutually exclusive events, then the probability of the event E_1 and E_2 is 0:

$$P(E_1 \cap E_2) = 0$$

If E_1 and E_2 are not mutually exclusive, then the probability of $E_1 \cap E_2$ is the number of elements common to E_1 and E_2 divided by the number of elements in the sample space.

For example, suppose we roll two dice. The sample space S contains thirty-six elements. Let E_1 represent the event "the sum is greater than 9," and E_2 represent the event "the sum is an odd number greater than 8." We have

$$E_1 = \{(46), (55), (64), (56), (65), (66)\}$$
$$E_2 = \{(36), (45), (54), (63), (56), (65)\}$$

The set $E_1 \cap E_2$ of elements common to E_1 and E_2 is

$$E_1 \cap E_2 = \{(56), (65)\}$$

Then

$$P(E_1 \cap E_2) = \tfrac{2}{36} = \tfrac{1}{18}$$

The set of elements in E_1 or E_2 (here "or" is used in the inclusive sense*) is denoted by $E_1 \cup E_2$, which is read "the union of E_1 and E_2," or simply "the event E_1 or E_2." It can be proved, although we shall not prove it here, that for any two events of a sample space

(1) $$P(E_1 \cup E_2) = P(E_1) + P(E_2) - P(E_1 \cap E_2)$$

If E_1 and E_2 are mutually exclusive events, then $P(E_1 \cap E_2) = 0$, and formula (1) reduces to

(2) $$P(E_1 \cup E_2) = P(E_1) + P(E_2)$$

We shall illustrate the truth of these formulas by examples.

Suppose we roll a pair of dice. What is the probability of the event "a sum of 7 or a sum of 11"? Here E_1 is the event "a sum of 7" and E_2 is the event "a sum of 11." Then

$$E_1 = \{(16), (25), (34), (43), (52), (61)\}$$
$$E_2 = \{(56), (65)\}$$

* See Chapter 1, Section 5.

Notice that E_1 and E_2 are mutually exclusive events. Now

$$E_1 \cup E_2 = \{(16), (25), (34), (43), (52), (61), (56), (65)\}$$

The sample space is the familiar sample space with thirty-six elements. Hence

$$P(E_1) = \tfrac{6}{36}, \qquad P(E_2) = \tfrac{2}{36}$$

and

$$P(E_1 \cup E_2) = \tfrac{8}{36}$$

Hence

$$P(E_1 \cup E_2) = \tfrac{8}{36} = \tfrac{6}{36} + \tfrac{2}{36} = P(E_1) + P(E_2)$$

Now let us find the probability of the event "a sum of 7 or a 4 on one die." Then

$$E_1 = \{(16), (25), (34), (43), (52), (61)\}$$
$$E_2 = \{(14), (24), (34), (44), (54), (64), (41), (42), (43), (45), (46)\}$$
$$E_1 \cap E_2 = \{(34), (43)\}$$
$$E_1 \cup E_2 = \left\{ \begin{array}{l} (16), (25), (34), (43), (52), (61), (14), (24), \\ (44), (54), (64), (41), (42), (45), (46) \end{array} \right\}$$

Then

$$P(E_1) = \tfrac{6}{36} = \tfrac{1}{6}$$
$$P(E_2) = \tfrac{11}{36}$$
$$P(E_1 \cap E_2) = \tfrac{2}{36}$$

We see that

$$P(E_1 \cup E_2) = P(E_1) + P(E_2) - P(E_1 \cap E_2) = \tfrac{6}{36} + \tfrac{11}{36} - \tfrac{2}{36} = \tfrac{15}{36}$$

EXERCISE 3

1. Which of the pairs of events listed below are mutually exclusive?
   (a) In tossing a coin, throwing heads, then tails.
   (b) In rolling a die, rolling an even number, rolling a 6.
   (c) Going upstairs, going downstairs.
   (d) In drawing a single card from a deck of cards, drawing a queen, drawing a heart.
   (e) In rolling a pair of dice, rolling a sum of 6, rolling a sum of 9.

2. What is the probability of each of the following events when rolling two dice?
   (a) A sum of 3 or a sum of 2.
   (b) A sum of 7 or a sum of 9.
   (c) A sum of 8 or a sum of 6.
   (d) A sum of 10 or a sum of 4.
   (e) A sum of 12 or a sum of 3.

3. In a box are four green, five yellow, and two red beads. All the beads are the same size and shape. One bead is picked at random.
   (a) What is the probability of picking a yellow bead?
   (b) What is the probability of picking a red bead?
   (c) What is the probability of picking a green or a red bead?
   (d) What is the probability of picking a yellow or a green bead?
   (e) What is the probability of picking a black bead?

4. During the past four years an instructor has taught 1000 students. In this period he has given A as a final grade to 70 students and B as a final grade to 320 students. Based on this data, what is the probability that a student selected at random will receive an A or a B?

5. The names of four Yorkshire terriers, twelve poodles, eight shelties, and five scotties are placed in a box at a pet shop. Each dog is owned by a different person. A collar is awarded by drawing at random a name from the box.
   (a) What is the probability that the winner will be a poodle?
   (b) What is the probability that the winner will be a Yorkshire terrier?
   (c) What is the probability that the winner will be either a Yorkshire terrier or a poodle?
   (d) What is the probability that the winner will be a sheltie or a scottie?

6. A bag has been filled with 100 jelly beans of assorted colors: there are thirty red, twenty-five yellow, twenty green, twenty pink, and five black. The jelly beans are thoroughly mixed so that the chance of getting any one jelly bean is as likely as another. If you pick one bean from the bag, what is the probability you get:
   (a) black
   (b) red or green
   (c) pink or red.

7. Two coins are tossed. If E_1 is the event "tails on the first coin" and E_2 is the event "the coins match," what is the probability of tails on the first coin or the coins match?

8. Two dice, one red and one green, are rolled. What is the probability of a sum of 11 or a numeral different from 6 on the green die?
9. What is the probability of rolling a 2 or a 4 on a die?
10. An employer has three positions to fill. He decides to fill them by selecting at random from among five equally qualified applicants, A, B, C, D, and E. What is the probability that both A and B will be chosen or C and D will be chosen?
11. Let S be the set of sixty students studying at least one of the two languages, Spanish or Russian. If thirty students are studying Spanish, fifty students are studying Russian, and twenty students are studying Spanish and Russian, what is the probability that a student picked at random studies Spanish or Russian?

4. PROBABILITY OF INDEPENDENT EVENTS

Suppose we have three beads in a box: one red, one white, and one blue. If we draw a bead at random from the box, the probability of drawing a red bead is $\frac{1}{3}$. If we put the bead we drew back in the box and draw a second, the probability that the second bead is blue is $\frac{1}{3}$.

Now let us ask what is the probability of red on the first draw and blue on the second. The possible outcomes are

$$S = \{(R, R), (R, W), (R, B), (W, R), (W, W), (W, B), (B, R), (B, W), (B, B)\}$$

The event R, red on the first draw, is

$$R = \{(R, R), (R, W), (R, B)\}$$

Hence $P(R) = \frac{3}{9} = \frac{1}{3}$.

The event B, "blue on the second draw," is

$$B = \{(R, B), (W, B), (B, B)\}$$

Hence

$$P(B) = \frac{3}{9} = \frac{1}{3}$$

Looking at the possible outcomes we see that the event "red on the first draw *and* blue on the second draw," which we call $R \cap B$, is $\{(R, B)\}$. Hence

$$P(R \cap B) = \frac{1}{9}$$

Notice that

$$P(R \cap B) = P(R) \cdot P(B)$$

Now let us use the same box, but this time draw a bead from the box and then draw another bead from the box without replacing the first bead. We ask ourselves: What is the probability of red on the first draw and blue on the second draw? We can draw a tree diagram of the possible outcomes (Figure 7.2).

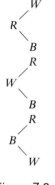

figure 7.2

A sample space for this experiment is

$$S = \{(R, W), (R, B), (W, R), (W, B), (B, R), (B, W)\}$$

Let E be the event "red on the first draw" and F be the event "blue on the second draw." The event $E \cap F$, "red on the first draw and blue on the second draw" is

$$E \cap F = \{(R, B)\}$$

and

$$P(E \cap F) = \tfrac{1}{6}$$

Notice that

$$E = \{(R, B), (R, W)\}$$

and

$$F = \{(R, B), (W, B)\}$$

Hence

$$P(E) = \tfrac{2}{6} = \tfrac{1}{3}$$

and

$$P(F) = \tfrac{2}{6} = \tfrac{1}{3}$$

We observe that in this case

$$P(E \cap F) \neq P(E) \cdot P(F)$$

In each preceding case we were drawing two beads from the box. In the first case, however, the first draw had no effect on the result of the second. When neither of two events depends on the other, the events are said to be **independent**. When two events X and Y are independent

$$P(X \cap Y) = P(X) \cdot P(Y)$$

In the second case, the second draw was affected by what happened on the first draw. In this case, we say the events are **dependent**. When two events are dependent

$$P(X \cap Y) \neq P(X) \cdot P(Y)$$

EXERCISE 4

1. If X, Y, and Z are independent events, then for the probability that all three occur, we have

$$P(X \cap Y \cap Z) = P(X) \cdot P(Y) \cdot P(Z)$$

   (a) State a similar property that holds for four independent events.
   (b) Find the probability of a head showing on each of four successive tosses of a coin.
2. Which of the following pairs of events are independent?
   (a) A 5 on the first roll of a die and a 2 on the second.
   (b) Heads on a toss of a penny and 1 on the toss of a die.
   (c) From a box containing three red, two black, and five white balls, first drawing a red ball, and then drawing a white ball in the second draw.
   (d) From a deck of cards drawing a king on the first draw and drawing an ace on the second draw.
   (e) Heads on the first toss of a penny and tails on the second.
3. What is the probability of drawing two kings from a deck of bridge cards if one card is drawn and then replaced before the second draw?
4. We have two spinners as shown below. Both pointers are made to spin.

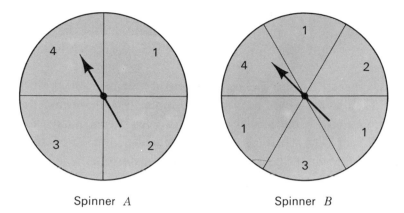

Spinner *A*   Spinner *B*

(a) What is the probability that both will stop on the region marked 1?

(b) What is the probability that both will stop on the region marked 2?

(c) What is the probability that the pointer on *A* will stop on the region marked 4 and the pointer on *B* will stop on the region marked 3?

(d) What is the probability that the pointer on *A* will stop on the region marked 1 and the pointer on *B* will stop on the region marked 4?

5. Ten cards are numbered 1 through 10, put in a box, and thoroughly mixed. Two cards are drawn from the box by a blindfolded person. What is the probability that:

(a) The numbers on both cards will be even?

(b) The sum of the two numbers is even?

(c) The sum of the two numbers is divisible by 5?

(d) The sum of the two numbers is less than 20?

6. On a baseball team, player *K* has a batting average of .300 and player *D* has a batting average of .250. Both players come to bat in the fifth inning.

(a) What is the probability that *K* and *D* get hits in the fifth inning?

(b) What is the probability that *K* or *D* or both get hits in the fifth inning?

5. MEASURES OF CENTRAL TENDENCY

The idea of **average** is familiar to everyone. No matter how little we know of mathematics, we are at home with batting averages, average salaries, average families, and the like. We realize that the purpose of an average is to represent a set of individual values or measures by one measure that is most representative of the members of the set. The idea of average is so handy that it is not surprising that several kinds of averages have been invented.

Averages are called **measures of central tendency**. The three measures of central tendency in common use are the **arithmetic mean**, the **median**, and the **mode**. All three measures are averages. Each has its advantages and limitations. The nature of the particular distribution of the data and the purpose for which the average is chosen determine which measure of central tendency is used.

6. THE ARITHMETIC MEAN

The **arithmetic mean**, usually referred to as the **mean**, of a set of numbers is obtained by adding the numbers and dividing the sum by the number of numbers in the set. If the symbols $X_1, X_2, X_3, \ldots, X_n$ indicate the numbers in a set of N numbers, the arithmetic mean, designated by $\overline{X}$, is given by

$$\overline{X} = \frac{X_1 + X_2 + \cdots + X_n}{N} = \frac{1}{N} \sum_{i=1}^{n} X_i$$

The symbol Σ, which is the Greek letter sigma, is used to indicate a sum. The symbol $\sum_{i=1}^{n} X_i$ is read "the sum for $i = 1$ to $i = n$ of X_i." Thus $\sum_{i=1}^{n} X_i$ means the sum $X_1 + X_2 + \cdots + X_n$.

Suppose the weights of five wrestlers to the nearest pound are: 204, 190, 240, 211, and 240. Then the mean weight of the wrestlers is 217 pounds.

$$\frac{204 + 190 + 240 + 211 + 240}{5} = \frac{1085}{5} = 217$$

If the numbers to be added in finding the mean are large, a considerable amount of work may be saved in the computation by averaging deviations from an **assumed mean**. We illustrate this method in computing the mean weight of the wrestlers. First let us assume a value of the mean, say 200 pounds. Next we find how much each weight differs from the assumed mean. In the example, these deviations are: $+4, -10, +40, +11$, and $+40$. We then obtain the mean of these deviations.

$$\frac{4 + (-10) + 40 + 11 + 40}{5} = \frac{85}{5} = 17$$

Adding this correction to the assumed mean we obtain the true mean: $200 + 17 = 217$.

If we designate the assumed mean as X_0 and the mean of the deviations as $\bar{Y}$, the foregoing method implies that the mean $\bar{X}$ is equal to the assumed mean plus the mean $\bar{Y}$. Then

$$\bar{X} = X_0 + \bar{Y}.$$

7. THE MEDIAN

The **median** of a set of N numbers is the middle number when they are arranged in sequential order. The median of the numbers

$$165, 185, 199, 242, 300, 307, 310$$

is 242.

If N is even, it is customary to take the median as the mean of the two middle numbers. The median of

$$126, 134, 157, 199, 199, 204$$

is the mean of 157 and 199.

$$\frac{157 + 199}{2} = \frac{356}{2}$$

$$= 178$$

8. THE MODE

The **mode** of a set of N numbers is the number that occurs most frequently. It is possible for a set of numbers to have more than one mode. If two or more numbers occur the maximum number of times, each number is called a mode. The mode of

$$18, 19, 24, 24, 27, 29, 31, 31, 31, 37, 39$$

is 31.

The set

$$26, 27, 29, 29, 29, 34, 37, 37, 37, 39, 42$$

has two modes, 29 and 37.

EXERCISE 5

1. Find the mean, median, and mode of each of the following sets.
   (a) 2, 3, 4, 7, 5, 3, 3, 10, 4, 9.
   (b) 24, 21, 20, 25, 21, 27.
   (c) 3, 3, 1, 2, 2, 3, 5, 4, 2, 3, 4, 4, 2, 4, 3, 3.
2. Find the mean of the numbers 1 through 10.
3. Find the mean of 1461, 1464, 1460, 1467, 1470, 1462. Use the assumed mean 1465.
4. Find the mean of 14.22, 14.25, 14.56, 14.77, 14.11, 14.32. Use the assumed mean 14.25.
5. The mean scores in two sections of a mathematics class were 84 and 92. Find the mean score of the two classes.
6. The barrels of crude oil produced daily in four of the United States is given below. What is the mean number of barrels of crude oil produced in these four states?

Texas	54,400
Oklahoma	23,300
Louisiana	12,400
New Mexico	7,800

7. Listed below are the annual salaries in a small retail store.

Manager	$14,000
Assistant manager	12,000
Buyer	10,000
Bookkeeper	6,000
Sales clerk	5,000
Stock clerk	3,000
Janitor	2,500

   (a) Find the mean salary of the store employees.
   (b) Find the median salary of the store employees.
8. The daily output of fifteen factory workers on a piece-work job was: 111, 119, 116, 128, 115, 140, 121, 128, 116, 118, 128, 130, 131, 133, and 131.
   (a) What is the mean number of pieces produced?
   (b) What is the median number of pieces produced?
   (c) What is the mode number of pieces produced?
9. If $\overline{X}$ is the mean of $X_1, X_2, \ldots, X_n$, what is the mean of:
   (a) $cX_1, cX_2, \ldots, cX_n$?
   (b) $X_1 + k, X_2 + k, \ldots, X_n + k$?

9. COMPARING THE THREE AVERAGES

The most frequently used measure of central tendency is the arithmetic mean. It is easy to compute and define, takes all the measures into consideration, and is well defined for algebraic manipulation. The mean is a magnitude average. It is the value each measure would have if all were equal. One of the chief advantages of the mean is its reliability in sampling. The mean is not typical if the data include a few extreme values in one direction.

The median is a positional average. The number of measures greater than the median is the same as the number that is less. The median, although easy to calculate and define, is not influenced by extreme measures. If measures are concentrated in distinct and widely separated groups, the median may be of little value as a measure of central tendency.

The mode is thought to be the most typical measure of all (since it is the most frequently occurring), but it does not take account of the other values in the data. It is less important than the median because of its ambiguity.

EXERCISE 6

1. A drugstore owner sold the following five brands of shampoo. In reordering would the druggist be interested in mean, median, or mode? Why?

Brand	Number of Bottles Sold
A	4
B	6
C	30
D	12
E	15

2. The following wages were paid by a factory:

Annual Wage	Number Receiving This Wage
$50,000	1
25,000	1
15,000	2
10,000	5
8,000	12
5,000	40
3,000	4

(a) What is the mean wage?

(b) What is the median wage?

(c) What is the mode wage?

(d) Which average would a union leader use in presenting arguments for salary increases?

(e) Which average would the factory owner use in advertising for help?

(f) Which average would the Internal Revenue Bureau use in speaking of the taxable income?

3. What average would be used in the following?

(a) In meteorology, for obtaining the average rainfall.

(b) In the garment industry, for determining the number of shirts of each size to manufacture.

(c) In business, for computing average wages.

(d) In education, for determining the average age of sixth grade children.

4. Suppose you are a buyer for a department store. Which type of average will probably be of the greatest value to you?

5. The contributions by alumni of a small college to a building fund are given below.

Amount in Dollars	Number of Contributions
1,000	1
500	2
100	8
50	12
25	20
15	5
10	7

(a) Compute the three averages for this data.

(b) What is the "usual" contribution?

(c) What average would best represent this data?

10. MEASURES OF VARIABILITY

The grades for two tests given to a mathematics class were the following:

Test 1	Test 2
100	80
90	75
80	75
75	70
75	70
70	70
70	70
70	70
65	65
65	65
65	65
55	60
40	60
40	55
40	55

The mean grade in Test 1 is 66.7 which we round to 67. On Test 2 the mean is again 67. If we knew only the mean we probably would assume that the class did the same on each test. Looking at the grades on the two tests, however, we see that this is not the case. There are both higher and lower grades on the first test. There is less variability on the second test. If each student had received a grade of 67 on the second test, the instructor would certainly not consider this performance of the class comparable to the performance of the class on the first test. We need, in addition to the measures of central tendency, a statistical measure to indicate the extent to which the variates tend to spread out. Such measures are called **measures of variability**.

The simplest measure of variability is the **range**. It tells the difference between the highest and the lowest measures. The range for Test 1 is $100 - 40 = 60$. The range for Test 2 is $80 - 50 = 30$. This indicates more variability on Test 1.

The range is a poor measure of variability because it depends only on two measures, telling us nothing about the remaining measures. Because the range has shortcomings as a measure of variability, others have been devised. One of them is the standard deviation denoted by the Greek letter σ.

To see intuitively how the standard deviation was derived, let us consider deviation from the mean by examining three sets of data:

Set 1: 15, 15, 15, 15, 15, 15
Set 2: 12, 12, 12, 18, 18, 18
Set 3: 10, 12, 14, 16, 18, 20

In each case, the mean is 15. What does this average mean? We might predict any number chosen at random from the set will not deviate much from 15. In fact, the errors we incur in selecting numbers greater than 15 are compensated by the errors in selecting numbers less than 15. For example, in Set 3,

10 undershoots 15 by 5	$10 - 15 = -5$
12 undershoots 15 by 3	$12 - 15 = -3$
14 undershoots 15 by 1	$14 - 15 = -1$
16 overshoots 15 by 1	$16 - 15 = 1$
18 overshoots 15 by 3	$18 - 15 = 3$
20 overshoots 15 by 5	$20 - 15 = 5$
	Total deviations $= 0$

If we add all the deviations, we find that the total is 0. Similarly, using 15 as a predicted value, the sum of the deviations for each of the other two sets is 0. But this does not say anything about how "good" the prediction is. Looking at the data, we can see that using 15 as a prediction for Set 1 is excellent since each statistic in the set is exactly 15. Again, from the data, we see that 15 is not too meaningful for Set 2 or Set 3, since at no time do we actually get the number 15 in these sets.

We need a measure that will give us an idea of how much deviation is involved without having these deviations "cancel" out each other because one deviation is positive and another negative. We can avoid this by considering the "absolute distance" (absolute value of the distance) of each deviation. Since absolute values are more difficult to work with than squares, and since squaring deviations is as likely to give an accurate picture of the deviations provided we take into account the exaggerated magnitudes in squaring, we can turn to squaring each deviation from the mean (thus we have no negative deviations) and taking the arithmetic mean of these squares. We have essentially what is called the **standard deviation**. It is the square root of the mean of the squares of the deviation of each score from the mean:

$$\sigma = \sqrt{\frac{1}{N} \sum_{i=1}^{n} (X_i - \bar{X})^2}$$

To illustrate let us find the standard deviation for Test 1 in Table 7.2 and for Test 2 in Table 7.3.

TABLE 7.2

Score	Deviation from Mean (66.7)	Deviation Squared
100	33.3 (100 − 66.7)	1108.9
90	23.3 (90 − 66.7)	542.9
80	13.3	176.9
75	8.3	68.9
75	8.3	68.9
70	3.3	10.9
70	3.3	10.9
70	3.3	10.9
65	− 1.7	2.9
65	− 1.7	2.9
65	− 1.7	2.9
55	−11.7	136.9
40	− 26.7	712.9
40	− 26.7	712.9
40	− 26.7	712.9
$N = 15$		4283.5

Substituting the values from Table 7.2 in the standard deviation formula, we have

$$\sigma = \sqrt{\frac{4286.5}{15}} = 16.9$$

TABLE 7.3

Score	Deviation from Mean (67)	Deviation Squared
80	13	169
75	8	64
75	8	64
70	3	9
70	3	9
70	3	9
70	3	9
70	3	9
65	−2	4
65	−2	4
65	−2	4
60	−7	49
60	−7	49
55	−12	144
55	−12	144
$N = 15$		740

The values from Table 7.3 give

$$\sigma = \sqrt{\frac{740}{15}} = 7.0$$

These two results indicate that Test 1, with the higher standard deviation, gives more variable results than Test 2. That is, the scores on Test 1 have a greater tendency to diverge from the mean score.

In general, *a relatively small standard deviation indicates that the measures tend to cluster close to the mean, and a relatively high one shows that the measures are widely scattered from the mean.*

Now that we know how to calculate the standard deviation, let us see what use we can make of it. Suppose the mean score on a mathematics test is 42 and the standard deviation is 6 points. If you have a score of 48 you are 6 points above the mean, or as many points as one standard deviation above the mean. Suppose that the mean on a second test is 50 and the standard deviation is 10. If you score 60 on the test, you would be 10 points, or one standard deviation above the mean. Since you are one standard deviation above the mean on each test, you know that you did equally well on both tests.

EXERCISE 7

1. Compute the standard deviation for the following scores.
   (a) 16, 24, 21, 17, 10, 19, 23, 20, 15, 12
   (b) 17, 16, 15, 14, 13, 12, 11
2. On a mathematics test the following scores were made in a class of fifteen students. Find the mean, mode, median, range, and standard deviation of these scores: 98, 65, 80, 75, 88, 95, 90, 40, 82, 60, 78, 90, 80, 70, 50
3. A machine turns out three-inch bolts. Samples taken during the day were measured with a micrometer with the following results: 3.03, 3.05, 2.99, 3.02, 3.06, 2.95, 2.98, 3.08, 3.00, 3.02, 2.94, 2.99. Find the mean and standard deviation of this sample.
4. Two tests were given. On the first the mean was 80 and the standard deviation was 2. On the second test the mean was 75 and the standard deviation was 4. On which of the two tests did the following students do better?

Student	First Test Score	Second Test Score
(a) Mark	80	80
(b) Jane	40	50
(c) Kay	50	18
(d) Leo	95	90
(e) Deane	76	82

5. Suppose you are 10 pounds heavier than the average person of your age. Suppose the average weight of persons of your age is 130 pounds and the standard deviation is 5 pounds. How many standard deviations is your weight above the mean?

6. Two chemistry tests were given to six students. On the first test the mean was 54 and the standard deviation was 8. On the second test the mean was 28 and the standard deviation 10.

Student	First Score	Second Score
Jan	62	38
Lela	66	23
Fred	46	28
Arthur	54	33
Gayle	70	43
Walter	50	18

(a) Which students improved in regard to the second test?
(b) Which students improved most in regard to the second test?
(c) Which students came down on the second test?
(d) Which students did equally well on both tests?

OSWALD VEBLEN (1880–1960), left, one of the America's leading mathematicians, was born in Iowa. He studied at Harvard and the University of Chicago. He taught at the University of Chicago and at Princeton, and in 1932 he became professor at the Princeton Institute for Advanced Study. Veblen's most notable contributions involved differential and projective geometry, the foundations of geometry and topology. Finite geometries were not brought into prominence until 1906 when Veblen and W. H. Bussey made a study of them. Dr. Robert Oppenheimer is shown on the right.

Finite Geometry/8

1. WHAT IS FINITE GEOMETRY?

The systems of numbers we usually work with—whole numbers, integers, rational numbers—involve infinitely many elements in the set. However, we have also worked with number systems, such as modular systems,* which have only a finite number of elements. The geometry studied in high school deals with infinite sets of points, lines, and planes. Did you ever think of a geometry with only a finite number of points and lines? **Finite geometry** is just such a geometry.

It was G. Fano who, in 1892, first considered a finite geometry, a three-dimensional geometry containing fifteen points, thirty-five lines, and fifteen planes, each plane containing seven points and seven lines. Finite geometries were not brought into prominence until 1906, when O. Veblen and W. H. Bussey made a study of them. Since that time, the study of finite geometries has grown considerably.

In the study of finite geometry one of the most difficult things we have to remember is *not* to read a meaning into the undefined terms. For instance, a line as we know it from Euclidean geometry has many properties: it has continuity, that is, it is continuous with no holes in it; it is composed of an infinite set of points; unless otherwise stated, it is assumed to be straight.

In the study of finite geometry we must forget these properties of a line and think of it as only a collection of points. In reading the postulates,

* See Chapter 4.

definitions, and theorems of finite geometry, we must keep in mind that whenever two or more points or lines are mentioned, it will be understood that we mean *distinct* points and lines.

2. SIX-POINT GEOMETRY

In **six-point geometry** we shall be concerned with a set S of undefined elements called "points" and "lines." We must be careful not to read any more meaning into these terms than is justified by the postulates.

The postulates of six-point geometry are:

POSTULATE 1: Each pair of lines in S has at least one point in common.

POSTULATE 2: Each pair of lines in S has not more than one point in common.

POSTULATE 3: Every point in S is on at least two lines.

POSTULATE 4: Every point in S is on not more than two lines.

POSTULATE 5: The total number of lines in S is four.

These postulates were selected arbitrarily. We could have had an entirely different set if we had so desired. There are some rules that must be observed in constructing a set of postulates for a mathematical system such as a finite geometry. The most important rule is: it must be possible for all the postulates to be true at the same time. This is called **consistency**. If it is possible for all the postulates to be true at the same time, then the set of postulates is said to be **consistent**.

Let us look at our set of postulates and determine whether or not they are consistent. There are several ways to do this, but one of the easiest is to see if we can set up a model. What do we mean by a model? A **model** simply means an example of some kind in which all the postulates are true. A model is sometimes called an **interpretation**. We will now give an interpretation of the undefined terms in our postulates. The postulates as they are written are *neither* true nor false, but when we assign meanings to the undefined terms then the postulates are *either* true or false. If we can find just one interpretation of the undefined terms such that all of the postulates are true, then we have shown that it is possible for all to be true at the same time and so be consistent.

3. A MODEL FOR SIX-POINT GEOMETRY

Can we find a model for the five postulates of the six-point geometry? Let us think for a moment of the set S as a set of tin soldiers arranged in rows. Then the soldiers represent the points and the rows represent the lines. We have now given an interpretation or meaning to our undefined terms. Remember that the lines (rows) are just a collection of points (soldiers). They are completely empty between the points (soldiers).

How can we decide how many tin soldiers we will need? Since we wish to make our postulates true, we must have one and only one soldier common to each pair of rows (Postulates 1 and 2). We must have every soldier in two and only two rows (Postulates 3 and 4), and we must have four rows of soldiers (Postulate 5). This means that the number of soldiers needed will be the same as the number of distinct pairs of things (rows) that can be chosen from four things (rows). Suppose the rows are represented by I, II, III, and IV. Then the number of distinct pairs that can be made from these four rows are

I-II	I-III	I-IV
II-III	II-IV	III-IV

Each distinct pair of rows will have only one soldier in common. This tells us that we will need exactly six tin soldiers arranged in four rows. For convenience let us name the six soldiers A, B, C, D, E, and F. The soldier in rows I and II we shall call A; the soldier in rows I and III, B; the soldier in rows I and IV, C; the soldier in rows II and III, D; the soldier in rows II and IV, E; and the soldier in rows III and IV, F. Then

I-II	A
I-III	B
I-IV	C
II-III	D
II-IV	E
III-IV	F

Since we are to use each soldier in two rows, if we count the number of times we could use the six soldiers, we find it is 2 times 6, or 12.

We have four rows, so we see that we will have three soldiers in each row. Let us see if we can now draw a diagram showing how the soldiers might be placed. Figure 8.1 shows such a diagram. Notice that the postulates did not state that a line must be straight, nor did they state that a line must not be straight, consequently we may have our lines curved or straight. The two diagrams in Figure 8.1 illustrate these two

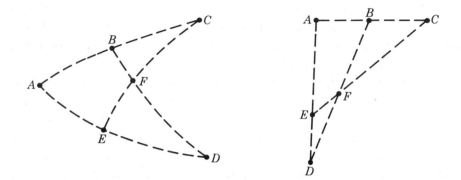

figure 8.1

possibilities. We must remember that our lines are not like ordinary line segments. They are empty between points. The broken lines in the diagram are only for our convenience in seeing the rows of which we are speaking.

We could name the rows in either of the two diagrams by listing the soldiers in the rows. In either of the two diagrams, let us call the row consisting of *ABC*, row I; the row consisting of *AED*, row II; the row consisting of *BFD*, row III; and the row consisting of *CFE*, row IV. We now have the rows and the soldiers in them as follows:

Row I is *ABC*
Row II is *AED*
Row III is *BFD*
Row IV is *CFE*

We must now go back to the postulates and see whether they are all true in our arrangement of tin soldiers in rows. Postulate 1 says that each pair of rows must have at least one soldier in common. Rows I and II have soldier *A* in common; rows I and III, soldier *B*; rows I and IV, soldier *C*; rows II and III, soldier *D*; rows II and IV, soldier *E*; and rows III and IV, soldier *F*. No pair of rows has more than one soldier in common, hence Postulate 2 is satisfied. Looking at the diagram we see that every soldier is in at least two rows, satisfying Postulate 3, and that no soldier is in more than two rows, satisfying Postulate 4. There are four rows in all, which satisfies Postulate 5.

EXERCISE 1

1. (a) Translate the postulates of six-point geometry into statements about beads and wires, interpreting "point" as bead, and "line" as wire.
   (b) Verify that the bead-wire interpretation satisfies all the postulates of six-point geometry.
2. (a) Translate the postulates of six-point geometry into statements about members and committees, interpreting "point" as member, and "line" as committee.
   (b) Verify that the member-committee interpretation satisfies all the postulates of six-point geometry.
3. (a) Translate the postulates in six-point geometry into statements about trees and rows of trees in an orchard, interpreting "point" as tree, and "line" as row.
   (b) Verify that the tree-row interpretation satisfies all the postulates of six-point geometry.
4. (a) Translate the postulates of six-point geometry into statements about stars and galaxies, interpreting "point" as star, and "line" as galaxy.
   (b) Verify that the star-galaxy interpretation satisfies all the postulates of six-point geometry.

4. THEOREMS OF SIX-POINT GEOMETRY

Just what have we shown by setting up this model? We have shown that in at least one instance it is possible for all five of the postulates to be true at the same time. We can now say that the five postulates are consistent.

Now that we have set up our postulates and have shown that they are consistent, what can we do with them? Can we find any other statements that would be true if the postulates are true? If so, these statements are called **theorems**.

When we set up our model of tin soldiers in rows, we discovered two statements that were not mentioned in the postulates. The first statement we found was that there are exactly six points in S. Let us state this fact as Theorem 8.1 and see if it would always be true if the postulates were true.

THEOREM 8.1: **There will be exactly six points in S.**

> *Proof:* There are exactly four lines in S. Let us   Postulate 5
> call them I, II, III, and IV.
>
>   Each pair of these lines has at least   Postulate 1
> one point of S in common. Let us call
> these points I-II, I-III, I-IV, II-III, II-IV,
> and III-IV.
>
>   These points are the only possible   Postulate 2
> points.
>
>   No two of these points coincide.   Postulate 4
>
>   Therefore there are exactly six points
> in S.

The other statement which we discovered was that there are exactly three points on each line in S. Let us state this statement as Theorem 8.2.

THEOREM 8.2: **There are exactly three points on each line in S.**

> *Proof:* For the proof of this theorem we can look at the points listed in the proof of Theorem 8.1 and actually count the points on each line. On line I we have the points I-II, I-III, I-IV; on line II we have the points I-II, II-III, II-IV; on line III we have the points I-III, II-III, and III-IV; on line IV we have the points I-IV, II-IV, and III-IV. Therefore, since this list contains all possible points in S and no other points, there must be exactly three points on each line in S.

Since we have arbitrarily set up these postulates, we can also arbitrarily make up **definitions**. Let us define **parallel points** in the following manner:

DEFINITION 8.1: **Two points that have no lines in common are called parallel points.**

If we look carefully at the diagram in Figure 8.1 we see that each point in S has one and only one point parallel to it. We will state this as Theorem 8.3.

THEOREM 8.3: Each point in S has one and only one point parallel to it.

Proof: Any given point in S is determined by exactly two lines in S.	Postulates 1 and 2
There remain exactly two lines in S on which this point does not lie.	Postulate 5
There must be one and only one point in S determined by the remaining two lines in S. Therefore there is one and only one point in S parallel to any given point in S.	Postulates 1 and 2

In looking closely at Theorem 8.3 we discover another statement that follows from the same proof. Since this statement is closely related to Theorem 8.3, we call it a corollary to Theorem 8.3.

COROLLARY 8.1: On a line in S, not containing a given point in S, there is one and only one point parallel to a given point.

The proof of this corollary is left to the reader.

If we interchange the word "point" with the word "line" and the words "containing a" with the words "lying on" in the foregoing corollary, we obtain the statement: "On a point in S, not lying on a given line in S, there is one and only one line parallel to a given line." This statement should be familiar to those who have studied plane geometry.

Notice the following pairs of statements:

Two points determine a line.
Two lines determine a point (that is, intersect in).

It will be observed that if we interchange the words "point" and "line" in these statements, each becomes the other. This principle is known as the **principle of duality**, and it has very important consequences. If we can show complete duality between two sets of postulates, then we can assume all of the theorems that have been proved for one set of postulates to be automatically true for the second set with no need to prove them individually. We can see that in any system containing a great many theorems this would save a great deal of work.

EXERCISE 2

1. Referring back to the model in problem 2 of Exercise 1, where the points of S were represented by members and the lines by committees, name the club members as follows: May, James, Sue, Tom, Bob, and Alice.
   (a) List the members of the four committees as they might be composed.
   (b) Check to see that all five postulates are satisfied.
   (c) If May is represented by E in Figure 8.1 and she refuses to serve on the same committee as Alice, identify Alice in this diagram. How are these points related in six-point geometry?
   (d) List the members who are "parallel."
2. Set up another model for the five postulates giving your own interpretation to the undefined terms. Rewrite the five postulates putting your interpretation in place of the words "point" and "line" and making other changes necessary for correct English.
3. Referring to Figure 8.1, list the six points from A to F and tell which point is parallel to it.
4. Rewrite the five postulates of six-point geometry exchanging the words "point" and "line" in each of their occurrences and replacing the word "have" with the words "lie on" and the words "is on" with the words "goes through." Below is a diagram for the postulates that result from the replacement.

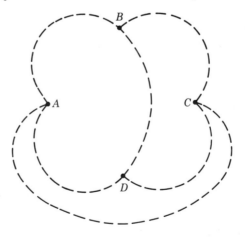

5. Draw another diagram that could represent the postulates of problem 4.
6. Rewrite the theorems, definition, and corollary as given in Section 4 to comply with the postulates in problem 4. (That is, write the duals of these statements.)

7. Which of the postulates of problem 4 would not be true in the following figure?

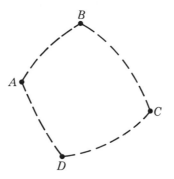

8. How many lines must there be in a model for the postulates in problem 4?
9. Is the following figure a correct model for the five postulates of six-point geometry? If not, which of the postulates are not satisfied?

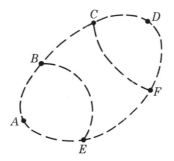

10. Interpreting the points in Figure 8.1 as San Diego, Los Angeles, Oakland, San Francisco, Portland, and Seattle, construct a permissible interpretation of six-point geometry by choosing appropriate airline flights.
11. Letting the cities of San Diego, Los Angeles, Oakland, and San Francisco represent the points in the diagram in problem 4, choose flights to represent the lines. List the pairs of parallel routes.

5. INDEPENDENCE AND COMPLETENESS

To create a finite geometry we have only to select some **undefined terms** and some unproved statements, called **postulates**, about the undefined terms. We then define new terms in terms of the original undefined terms. Next we prove the truth of new statements, called **theorems**, from the postulates. It is essential that our postulates be consistent.

Another property of a set of postulates for a finite geometry which is desirable, although not logically necessary, is **independence**. We say a set of postulates is **independent** if no one postulate can be deduced from the remaining ones. If any one of the postulates can be deduced from the remaining ones it is said to be **dependent**. In this case, there is no need to include this statement among the postulates in the finite geometry. This simply means that we say nothing in the postulates except what is necessary. Any statements that are true because of some of the other postulates, we prove as theorems rather than stating them as additional postulates.

To prove the independence of any postulate we only have to show that the other postulates of the system, together with some contradiction of the postulate in question, form a consistent set of postulates. Returning to the postulates of six-point geometry it is not difficult to supply independence proofs for each of the postulates. For convenience, let us restate the postulates.

POSTULATE 1: Each pair of lines in S has at least one point in common.

POSTULATE 2: Each pair of lines in S has not more than one point in common.

POSTULATE 3: Every point in S is on at least two lines.

POSTULATE 4: Every point in S is on not more than two lines.

POSTULATE 5: The total number of lines in S is four.

To show the independence of Postulate 1, let us interpret S to consist of five points, *A*, *B*, *C*, *D*, and *E*, distributed among four lines, I, II, III, and IV, as follows:

I	*AB*
II	*ACD*
III	*BDE*
IV	*CE*

In this interpretation Postulate 1 fails because pairs of lines I and IV have no point in common. On the other hand, no pair of lines has more than one point in common, so Postulate 2 is satisfied. Each point is on at least two lines and not more than two lines, so Postulates 3 and 4 are satisfied. We have four lines in *S* so Postulate 5 is satisfied. In the same way, independence proofs can be supplied for Postulates 2, 3, 4, and 5.

The third property that we would like our set of postulates to have is **completeness**. A postulate system is said to be **complete** if it is always possible to prove every statement that can be made about the undefined terms of the system, or else to prove the contradiction of this same statement from the postulates of the system. It is not always possible to have a postulate system complete, but it is a desirable property to have if possible. It is somewhat easier to have completeness in a postulate system for a finite geometry than it is for a system that deals with an infinite number of points and lines or other undefined terms.

EXERCISE 3

1. Prove the independence of Postulate 2 of six-point geometry. (*Hint:* Interpret S to consist of seven points.)
2. Prove the independence of Postulate 4 of six-point geometry. (*Hint:* Interpret S to consist of four points.)
3. Prove the independence of Postulate 3 of six-point geometry. (*Hint:* Interpret S to consist of seven points.)
4. Prove the independence of Postulate 5 of six-point geometry.

6. SEVEN-POINT GEOMETRY

The finite geometry we studied in the previous sections of this chapter had only six points. We shall now state the postulates for a seven-point geometry. In seven-point geometry we will have a set S of undefined terms called "points" and "lines."

POSTULATE 1: If P_1 and P_2 are any two points in S, there is at least one line containing both P_1 and P_2.

POSTULATE 2: If P_1 and P_2 are any two points in S, there is at most one line containing both P_1 and P_2.

POSTULATE 3: If L_1 and L_2 are any two lines in S, there is at least one point that lies on both L_1 and L_2.

POSTULATE 4: There are exactly three points on each line.

POSTULATE 5: If L is any line in S, there is at least one point that does not lie on L.

POSTULATE 6: There exists at least one line.

Let us now construct a model to test our postulates for consistency. Let us think of the points as men and the lines as clubs to which they belong. There exists at least one club having exactly three members. Let us designate men by the symbols $A, B, C, D, \ldots$, and the club by ABC. There exists a man D who is not a member of club ABC. Now D must belong to the same club as A, and neither B nor C can belong to this club. Thus there exists another club ADE, and similarly a club BDF, where E and F are other men.

At present we have three clubs (ABC, ADE, BDF) and six men. Every pair of men, however, must belong to exactly one club. Man A now belongs to a club with B, C, D, and E. Thus A and F cannot belong to a club with B, C, D, or E. Hence there must be a seventh man in the club with A and F. And hence there is club AFG. If there were an eighth man H, then the club containing A and H could not have a member in common with BDF, since A is already in a club with B, D, and F. Thus the set consists of exactly seven men. By continuing this type of reasoning, clubs CEF, BEG, and CDG can be found. There are then exactly seven men and seven clubs, as indicated by the following seven columns:

$$A \quad A \quad B \quad A \quad C \quad B \quad C$$
$$B \quad D \quad D \quad F \quad E \quad E \quad D$$
$$C \quad E \quad F \quad G \quad F \quad G \quad G$$

With this interpretation it is easy to verify that each of the six postulates is fulfilled. Postulates 4, 5, and 6 obviously are justified. In order to verify Postulates 1 and 2, we must consider every possible pair of men and verify that there is one and only one club to which both men belong. For example, A and B belong to club ABC, but to no other club. The pairs and their common clubs are:

Club	Pairs of Men	Club	Pairs of Men
ABC	A, B	BDF	D, F
ABC	A, C	BEG	B, E
ABC	B, C	BEG	B, G
ADE	A, D	BEG	E, G
ADE	A, E	CDG	C, D
ADE	D, E	CDG	C, G
AFG	A, F	CDG	D, G
AFG	A, G	CEF	C, E
AFG	F, G	CEF	C, F
BDF	B, D	CEF	E, F
BDF	B, F		

Finally, to verify Postulate 3, we must consider every possible pair of clubs and verify that in each case there is at least one man who belongs to both clubs. The pairs of clubs and their common members are:

Member	Pairs of Clubs	Member	Pairs of Clubs
A	ABC, ADE	F	AFG, BDF
A	ABC, AFG	G	AFG, BEG
B	ABC, BDF	F	AFG, CDG
B	ABC, BEG	F	AFG, CEF
C	ABC, CDG	B	BDF, BEG
C	ABC, CEF	D	BDG, CDG
A	ADE, AFG	F	BDF, CEF
D	ADE, BDF	D	BEG, CDG
E	ADE, BEG	E	BEG, CEF
D	ADE, CDG	C	CDG, CEF
E	ADE, CEF		

We have now shown that our set of postulates is consistent by exhibiting a concrete model for the seven-point geometry.

EXERCISE 4

1. Draw a diagram that represents the seven-point geometry. Remember that the lines need not be straight. (*Hint*: One line may be a circle.)
2. Give another interpretation for the postulates of seven-point geometry. Verify that your interpretation satisfies all the postulates.
3. Consider the system consisting of seven triples of the form $P = (u, v, w)$. The seven points are:

$$P_1 = (1, 0, 0) \quad P_2 = (0, 1, 0) \quad P_3 = (0, 0, 1)$$
$$P_4 = (0, 1, 1) \quad P_5 = (1, 0, 1) \quad P_6 = (1, 1, 0)$$
$$P_7 = (1, 1, 1)$$

Let the L's be those sets of P's that satisfy the respective equations:

$$L_1 : u = 0 \qquad L_2 : v = 0 \qquad L_3 : w = 0$$
$$L_4 : v + w = 0 \qquad L_5 : u + w = 0 \qquad L_6 : u + v = 0$$
$$L_7 : u + v + w = 0$$

where all the calculations are computed modulo 2; that is, the results of all calculations are to be divided by 2 and only the remainders retained. List the P's that lie on each L. (For example, P_2 and P_3 and P_4 lie on L_1 because the first member (u) of P_2, P_3, and P_4 is 0.) Verify that in this system the postulates of the seven-point geometry are satisfied.

7. THEOREMS FOR SEVEN-POINT GEOMETRY

We can now prove some simple theorems that follow from the postulates of the seven-point geometry.

THEOREM 8.4: There exists at least one point.

Proof: By Postulate 6 there exists at least one line, and by Postulate 4 every line contains exactly three points. Hence the assertion that at least one point exists is surely true.

THEOREM 8.5: If L_1 and L_2 are any two lines, there is at most one point that lies on both L_1 and L_2.

Proof: To prove this theorem, let us assume the contrary* and suppose it is possible for two lines, say L_1 and L_2, to have two points, say P_1 and P_2, in common. This leads at once to a contradiction since Postulate 2 asserts that there is at most one line that contains each of two given points. Hence it follows that the two lines can have at most one point in common.

THEOREM 8.6: Two points determine exactly one line.

Proof: This follows immediately from Postulates 1 and 2.

THEOREM 8.7: Two lines have exactly one point in common.

Proof: This follows immediately from Postulate 3 and Theorem 8.5.

THEOREM 8.8: If P is any point, there is at least one line that does not contain P.

Proof: By Postulate 6 there exists at least one line. If this line, L, does not contain P, our proof is complete. Suppose

---

* This is an indirect proof. See Chapter 1, Section 13.

therefore, that L passes through P. By Postulate 4, L contains two points besides P. Call one of them P'. By Postulate 5 there is at least one point, say P'', that does not lie on L. By Theorem 8.6 there is a unique line, say L', that contains P' and P''. Moreover, L and L' are distinct, since L' contains P'' and L does not. Hence by Theorem 8.6, L and L' have exactly one point in common, and this point is P'. Therefore P, which lies on L, cannot lie on L'. In other words, L' is a line that does not contain P.

THEOREM 8.9: Every point lies on at least three lines.

Proof: Let P be an arbitrary point. By Theorem 8.8 there is at least one line, L, that does not pass through P, and by Postulate 4 this line contains three points, P_1, P_2, and P_3. By Theorem 8.6 each of these points determines with P a unique line. Moreover, all these lines are distinct, for if two of them coincided, that line would have two points in common with L, which is impossible by Theorem 8.7. Hence there are at least three lines passing through an arbitrary point P as asserted.

EXERCISE 5

1. State the duals of Theorems 8.6, 8.7, and 8.8.
2. Translate the postulates of the seven-point geometry into statements about politicians and committees. Interpret "point" as politician and "line" as committee. Verify that the politician-committee interpretations satisfy all of the postulates of the seven-point geometry.
3. Translate the theorems proved in Section 7 using politician for "point" and committee for "line." Change the English when necessary so that it reads correctly.
4. Prove: Any two points are on exactly one line.
5. Prove: There exist three points that are not on the same line.
6. Prove: Any two lines have exactly one point in common.

ARTHUR CAYLEY (1821–1895), left, and JAMES JOSEPH SYLVESTER (1814–1897), right, were close friends who inspired each other in their work on the theory of invariants and matrices. The two men were of very different character. Cayley was gentle, generous, and led a serene life. Sylvester, however, was hot tempered and spent much of his life "fighting the world." With two such different temperaments, the friendship was not always a happy one. Sylvester was often on the point of exploding but Cayley was the safety valve. Sylvester paid grateful tribute to Cayley in a lecture at Oxford in 1885 by saying, "Cayley, who, though younger than myself, is my spiritual progenitor—who first opened my eyes and purged them of dross so that they could see and accept the higher mysteries of our common Mathematical Faith." Cayley mentioned Sylvester frequently in his memoirs but never in such glowing terms as those used by Sylvester. In 1858 Cayley invented matrices and their algebra. Some sixty years later, in 1925, Heisenberg recognized in the algebra of matrices the tool he needed in his work on quantum mechanics.

Matrices/9

1. HISTORY

We are all familiar with sets on which binary operations* are defined. For example, we are familiar with the set W

$$W = \{0, 1, 2, 3, \ldots\}$$

of whole numbers and the operations of addition and multiplication. We are also familiar with the set of rational numbers and their operations. In Chapter 4 we became acquainted with the set

$$\{0, 1, 2, \ldots (m - 1)\}$$

and addition and multiplication modulo m.

In this chapter we shall study a mathematical system created in the nineteenth century by Arthur Cayley (1821–1895) and James Joseph Sylvester (1814–1897). The results of the discoveries of these men furnished a broad foundation upon which the modern theory of matrices was built.

In the early 1940's an upsurge of interest occurred in the study of matrix theory. This branch of mathematics is extremely useful not only to mathematicians but also to persons interested in biology, sociology, economics, engineering, physics, psychology, and statistics. We shall study this system as a matter of interest and also to introduce a new and powerful symbolism.

* See Chapter 6, Section 1.

2. THE FORM OF A MATRIX

A (real) **matrix** (plural: matrices) is a rectangular array of real numbers, written in the following form:

$$\begin{pmatrix} a_{11} & a_{12} \ldots a_{1n} \\ a_{21} & a_{22} \ldots a_{2n} \\ \vdots \\ a_{m1} & a_{m2} \ldots a_{mn} \end{pmatrix}$$

Here the letters a_{ij} stand for real numbers and m and n are positive integers. Observe that when we deal with such an array we put parentheses around it. Notice that m is the number of rows, and n is the number of columns of the matrix. The real numbers in the array are called the **elements** of the matrix. We call a matrix with m rows and n columns an m by n matrix. We write this $m \times n$. We call m and n the **dimensions** of the matrix. Figure 9.1 contains examples of matrices. Figure 9.1a is a 2×2 matrix; Figure 9.1b, a 3×3 matrix; and Figure 9.1c, a 3×2 matrix.

$$\begin{pmatrix} 1 & 2 \\ 3 & 4 \end{pmatrix} \qquad \begin{pmatrix} 1 & 7 & -3 \\ -2 & 0 & 6 \\ -7 & -8 & -5 \end{pmatrix} \qquad \begin{pmatrix} 1 & 2 \\ 3 & 8 \\ -1 & -7 \end{pmatrix}$$

$$(a) \qquad\qquad (b) \qquad\qquad\qquad (c)$$

figure 9.1

Matrices that have $m = n$ are called **square** matrices. We shall confine our study to 2×2 matrices, that is, matrices that have two rows and two columns.

We say that two $m \times n$ matrices are **equal** if and only if the elements in the corresponding positions are equal. This means that

$$\begin{pmatrix} a & b \\ c & d \end{pmatrix} = \begin{pmatrix} w & x \\ y & z \end{pmatrix}$$

if and only if $a = w$, $b = x$, $c = y$, and $d = z$. Thus

$$\begin{pmatrix} 1 & 2 \\ 3 & 4 \end{pmatrix} = \begin{pmatrix} 1 + 0 & 1 + 1 \\ 2 + 1 & 3 + 1 \end{pmatrix}$$

because $1 = 1 + 0$, $2 = 1 + 1$, $3 = 2 + 1$, and $4 = 3 + 1$.

EXERCISE 1

1. Which of the following pairs of matrices are equal?

(a) $\begin{pmatrix} 0 & 3 \\ 5 & 4 \end{pmatrix}$, $\begin{pmatrix} 0 & -3 \\ -5 & -4 \end{pmatrix}$

(b) $\begin{pmatrix} .75 & .5 \\ .2 & .1 \end{pmatrix}$, $\begin{pmatrix} \frac{3}{4} & \frac{1}{2} \\ \frac{1}{5} & \frac{1}{10} \end{pmatrix}$

(c) $\begin{pmatrix} \sqrt{9} & \sqrt{4} \\ \sqrt{16} & \sqrt{25} \end{pmatrix}$, $\begin{pmatrix} 3 & 2 \\ 4 & 5 \end{pmatrix}$

(d) $\begin{pmatrix} \sqrt{2} & 3 \\ 22 & 7 \end{pmatrix}$, $\begin{pmatrix} 3 & \sqrt{2} \\ 7 & 22 \end{pmatrix}$

2. Give the value of x that makes the following pairs of matrices equal.

(a) $\begin{pmatrix} 1 & 3 \\ 7 & 5 \end{pmatrix} = \begin{pmatrix} 1 & 3 \\ x & 5 \end{pmatrix}$

(b) $\begin{pmatrix} 7 & 2 \\ x+3 & 7 \end{pmatrix} = \begin{pmatrix} 7 & 2 \\ 9 & 7 \end{pmatrix}$

(c) $\begin{pmatrix} 3 & \frac{4}{5} \\ x+\frac{1}{2} & 5 \end{pmatrix} = \begin{pmatrix} 3 & \frac{4}{5} \\ \frac{5}{8} & 5 \end{pmatrix}$

(d) $\begin{pmatrix} \sqrt{9} & \sqrt{5} \\ 2\sqrt{6} & \sqrt{8} \end{pmatrix} = \begin{pmatrix} x+1 & \sqrt{5} \\ 2\sqrt{6} & \sqrt{8} \end{pmatrix}$

3. Give the dimensions of the following matrices.

(a) $\begin{pmatrix} 3 & 4 \\ 1 & 2 \\ 1 & 7 \end{pmatrix}$ (c) $(1 \quad 2 \quad 3 \quad 4)$

(b) $\begin{pmatrix} 8 & 1 & 0 & 3 \\ 4 & 2 & 7 & 9 \\ 8 & 3 & 6 & 7 \\ 4 & 2 & 1 & 1 \end{pmatrix}$ (d) $\begin{pmatrix} 2 & 3 & 7 & 6 \\ 1 & 0 & 0 & 1 \end{pmatrix}$

4. A certain automobile agency has its salesmen send in 3×5 matrices as sales reports where the rows stand, in order, for the

number of hardtops, sedans, and station wagons sold, and the columns, in order, for the colors black, blue, red, white, and green. It received reports from two salesmen as follows:

$$A: \begin{pmatrix} 2 & 3 & 4 & 1 & 0 \\ 3 & 3 & 1 & 1 & 4 \\ 1 & 1 & 5 & 6 & 2 \end{pmatrix} \qquad B: \begin{pmatrix} 4 & 1 & 0 & 0 & 0 \\ 3 & 2 & 3 & 5 & 1 \\ 7 & 4 & 1 & 0 & 4 \end{pmatrix}$$

(a) How many hardtops did A sell?
(b) How many green hardtops did B sell?
(c) How many sedans did B sell?
(d) Which man sold the most station wagons?
(e) How many cars did B sell?
(f) How many cars did A sell?
(g) How many white cars did B sell?
(h) How many red cars were sold by A and B?

5. Suppose Albert, Burton, and Cox go to the bookstore and purchase the following articles.

Albert: 2 notebooks, 4 textbooks, 2 football tickets.
Burton: 5 notebooks, 1 textbook, 1 football ticket.
Cox: 6 notebooks, 6 textbooks, 4 football tickets.

Write a 3 × 3 matrix whose rows give the various purchases.

6. The following matrix gives the vitamin content of three foods in chosen units. The rows represent, in order, foods I, II, and III. The columns represent, in order, vitamins A, B, C, and B_1.

$$\begin{pmatrix} .3 & .3 & 0 & 0 \\ .4 & 0 & .2 & .1 \\ .5 & .5 & .1 & .6 \end{pmatrix}$$

(a) How much vitamin B is in food I?
(b) How much vitamin B_1 is in food III?
(c) How much vitamin C would you get if you ate one unit of each of the three foods?

7. An instructor used the following matrix to indicate the grades received by five graduate students during a semester. The rows represent, in order, students I, II, III, IV, and V. The columns represent, in order, grades A, B, C, D, and F.

$$\begin{pmatrix} 2 & 3 & 1 & 0 & 0 \\ 4 & 1 & 1 & 0 & 0 \\ 3 & 1 & 1 & 1 & 0 \\ 5 & 0 & 0 & 0 & 1 \\ 1 & 0 & 3 & 1 & 1 \end{pmatrix}$$

(a) How many A's did the instructor give during the semester?
(b) How many A's and B's did student II receive?
(c) How many students received F's during the semester?
(d) How many C's did the instructor give during the semester?
(e) How many D's and F's did the instructor give during the semester?

3. ADDITION OF MATRICES

Matrices of the same dimensions are added elementwise. For example,

$$\begin{pmatrix} 2 & 3 \\ 4 & 5 \end{pmatrix} + \begin{pmatrix} 4 & 1 \\ 3 & 1 \end{pmatrix} = \begin{pmatrix} 2+4 & 3+1 \\ 4+3 & 5+1 \end{pmatrix} = \begin{pmatrix} 6 & 4 \\ 7 & 6 \end{pmatrix}$$

In general,

$$\begin{pmatrix} a & b \\ c & d \end{pmatrix} + \begin{pmatrix} e & f \\ g & h \end{pmatrix} = \begin{pmatrix} a+e & b+f \\ c+g & d+h \end{pmatrix}$$

Matrices of different sizes cannot be added.

If we wish to define addition of 2×2 matrices in words, we say that the sum of two 2×2 matrices is a 2×2 matrix whose elements are the sum of the elements in the corresponding places in the matrices to be added.

Since the numbers to be added are real numbers, the order in which we add them does not matter; the result is the same. Hence the order in which we add matrices does not matter. For example,

$$\begin{pmatrix} 2 & 3 \\ 4 & 5 \end{pmatrix} + \begin{pmatrix} 1 & 4 \\ 5 & 7 \end{pmatrix} = \begin{pmatrix} 3 & 7 \\ 9 & 12 \end{pmatrix}$$

$$\begin{pmatrix} 1 & 4 \\ 5 & 7 \end{pmatrix} + \begin{pmatrix} 2 & 3 \\ 4 & 5 \end{pmatrix} = \begin{pmatrix} 3 & 7 \\ 9 & 12 \end{pmatrix}$$

Because the order in which we add matrices does not change the sum, we say that matrix addition is **commutative**.

Let us consider a special 2×2 matrix in which every element is 0:

$$\begin{pmatrix} 0 & 0 \\ 0 & 0 \end{pmatrix}$$

Notice the result when this matrix is added to another 2×2 matrix. For example,

$$\begin{pmatrix} 2 & 4 \\ 1 & 6 \end{pmatrix} + \begin{pmatrix} 0 & 0 \\ 0 & 0 \end{pmatrix} = \begin{pmatrix} 2 & 4 \\ 1 & 6 \end{pmatrix}$$

$$\begin{pmatrix} a & b \\ c & d \end{pmatrix} + \begin{pmatrix} 0 & 0 \\ 0 & 0 \end{pmatrix} = \begin{pmatrix} a+0 & b+0 \\ c+0 & d+0 \end{pmatrix} = \begin{pmatrix} a & b \\ c & d \end{pmatrix}$$

A matrix with all elements equal to zero is called a **zero matrix**. The zero matrix is called the **additive identity** for a system of matrices.

Every real number a has an additive inverse $-a$ such that

$$a + (-a) = 0$$

Thus

$$2 + (-2) = 0$$
$$(-3) + -(-3) = 0$$
$$\tfrac{1}{2} + (-\tfrac{1}{2}) = 0$$

We now ask ourselves: Does every matrix have an additive inverse? Since every real number has an additive inverse and since the elements of matrices are real numbers, we are assured that every matrix has an **additive inverse**. Observe that

$$\begin{pmatrix} 2 & 3 \\ 4 & 6 \end{pmatrix} + \begin{pmatrix} -2 & -3 \\ -4 & -6 \end{pmatrix} = \begin{pmatrix} 2+(-2) & 3+(-3) \\ 4+(-4) & 6+(-6) \end{pmatrix} = \begin{pmatrix} 0 & 0 \\ 0 & 0 \end{pmatrix}$$

In general,

$$\begin{pmatrix} a & b \\ c & d \end{pmatrix} + \begin{pmatrix} -a & -b \\ -c & -d \end{pmatrix} = \begin{pmatrix} a+(-a) & b+(-b) \\ c+(-c) & d+(-d) \end{pmatrix} = \begin{pmatrix} 0 & 0 \\ 0 & 0 \end{pmatrix}$$

The **additive inverse** of

$$\begin{pmatrix} a & b \\ c & d \end{pmatrix}$$

is

$$\begin{pmatrix} -a & -b \\ -c & -d \end{pmatrix}$$

Notice the following additions:

1. $\left[\begin{pmatrix} 1 & 2 \\ 3 & 4 \end{pmatrix} + \begin{pmatrix} 3 & 0 \\ 1 & 2 \end{pmatrix}\right] + \begin{pmatrix} 2 & 1 \\ 1 & 1 \end{pmatrix} = \begin{pmatrix} 4 & 2 \\ 4 & 6 \end{pmatrix} + \begin{pmatrix} 2 & 1 \\ 1 & 1 \end{pmatrix} = \begin{pmatrix} 6 & 3 \\ 5 & 7 \end{pmatrix}$

$\begin{pmatrix} 1 & 2 \\ 3 & 4 \end{pmatrix} + \left[\begin{pmatrix} 3 & 0 \\ 1 & 2 \end{pmatrix} + \begin{pmatrix} 2 & 1 \\ 1 & 1 \end{pmatrix}\right] = \begin{pmatrix} 1 & 2 \\ 3 & 4 \end{pmatrix} + \begin{pmatrix} 5 & 1 \\ 2 & 3 \end{pmatrix} = \begin{pmatrix} 6 & 3 \\ 5 & 7 \end{pmatrix}$

Hence

$\left[\begin{pmatrix} 1 & 2 \\ 3 & 4 \end{pmatrix} + \begin{pmatrix} 3 & 0 \\ 1 & 2 \end{pmatrix}\right] + \begin{pmatrix} 2 & 1 \\ 1 & 1 \end{pmatrix} = \begin{pmatrix} 1 & 2 \\ 3 & 4 \end{pmatrix} + \left[\begin{pmatrix} 3 & 0 \\ 1 & 2 \end{pmatrix} + \begin{pmatrix} 2 & 1 \\ 1 & 1 \end{pmatrix}\right]$

2. $\left[\begin{pmatrix} -1 & 1 \\ -1 & -1 \end{pmatrix} + \begin{pmatrix} 2 & -1 \\ 3 & 0 \end{pmatrix}\right] + \begin{pmatrix} -2 & -1 \\ -1 & -2 \end{pmatrix} = \begin{pmatrix} 1 & 0 \\ 2 & -1 \end{pmatrix} + \begin{pmatrix} -2 & -1 \\ -1 & -2 \end{pmatrix}$

$= \begin{pmatrix} -1 & -1 \\ 1 & 3 \end{pmatrix}$

$\begin{pmatrix} -1 & 1 \\ -1 & -1 \end{pmatrix} + \left[\begin{pmatrix} 2 & -1 \\ 3 & 0 \end{pmatrix} + \begin{pmatrix} -2 & -1 \\ -1 & -2 \end{pmatrix}\right] = \begin{pmatrix} -1 & 1 \\ -1 & -1 \end{pmatrix} + \begin{pmatrix} 0 & -2 \\ 2 & -2 \end{pmatrix}$

$= \begin{pmatrix} 1 & 1 \\ 1 & -3 \end{pmatrix}$

Hence

$\left[\begin{pmatrix} -1 & 1 \\ -1 & -1 \end{pmatrix} + \begin{pmatrix} 2 & -1 \\ 3 & 0 \end{pmatrix}\right] + \begin{pmatrix} -2 & -1 \\ -1 & -2 \end{pmatrix}$

$= \begin{pmatrix} -1 & 1 \\ -1 & -1 \end{pmatrix} + \left[\begin{pmatrix} 2 & -1 \\ 3 & 0 \end{pmatrix} + \begin{pmatrix} -2 & -1 \\ -1 & -2 \end{pmatrix}\right]$

In general,

$\left[\begin{pmatrix} a & b \\ c & d \end{pmatrix} + \begin{pmatrix} e & f \\ g & h \end{pmatrix}\right] + \begin{pmatrix} w & x \\ y & z \end{pmatrix} = \begin{pmatrix} a & b \\ c & d \end{pmatrix} + \left[\begin{pmatrix} e & f \\ g & h \end{pmatrix} + \begin{pmatrix} w & x \\ y & z \end{pmatrix}\right]$

Since the order in which we group matrices in addition does not matter, we say that matrix addition is **associative**.

EXERCISE 2

1. Find the sums.

   (a) $\begin{pmatrix} 3 & -7 \\ 4 & 9 \end{pmatrix} + \begin{pmatrix} -5 & 3 \\ 7 & 9 \end{pmatrix}$

   (b) $\begin{pmatrix} 9 & 21 \\ 71 & -80 \end{pmatrix} + \begin{pmatrix} 49 & -37 \\ 101 & -87 \end{pmatrix}$

2. Find the sums.

   (a) $\begin{pmatrix} \frac{5}{6} & \frac{4}{5} \\ \frac{3}{8} & \frac{5}{6} \end{pmatrix} + \begin{pmatrix} \frac{1}{2} & \frac{3}{10} \\ \frac{1}{4} & \frac{7}{12} \end{pmatrix}$

   (b) $\begin{pmatrix} 1.9 & 3.7 \\ 8.4 & -1.6 \end{pmatrix} + \begin{pmatrix} -1.7 & 4.8 \\ -9.9 & -7.7 \end{pmatrix}$

3. Find the sums.

   (a) $\begin{pmatrix} \frac{3}{4} & \frac{5}{8} \\ -\frac{1}{2} & \frac{5}{7} \end{pmatrix} + \begin{pmatrix} \frac{3}{4} & \frac{7}{8} \\ \frac{5}{8} & \frac{4}{7} \end{pmatrix}$

   (b) $\begin{pmatrix} \sqrt{2} & 7 \\ -\sqrt{2} & 9 \end{pmatrix} + \begin{pmatrix} -\sqrt{2} & 8 \\ \sqrt{2} & 15 \end{pmatrix}$

4. Find the sums.

   (a) $\begin{pmatrix} -\sqrt{5} & -\sqrt{2} \\ 7 & 15 \end{pmatrix} + \begin{pmatrix} \sqrt{5} & \sqrt{2} \\ -7 & 27 \end{pmatrix}$

   (b) $\begin{pmatrix} \frac{1}{2} & \sqrt{3} \\ 5\sqrt{6} & \sqrt{8} \end{pmatrix} + \begin{pmatrix} \frac{3}{4} & 2\sqrt{3} \\ -4\sqrt{6} & -5\sqrt{8} \end{pmatrix}$

5. Give the replacements for x, y, z, and w that will result in true statements.

   (a) $\begin{pmatrix} x+3 & 1 \\ 3 & 9 \end{pmatrix} = \begin{pmatrix} 5 & 1 \\ 3 & z-3 \end{pmatrix}$

   (b) $\begin{pmatrix} \frac{x}{3}-1 & \frac{y}{4}+6 \\ -1 & -7 \end{pmatrix} = \begin{pmatrix} 0 & 4 \\ z+6 & w+14 \end{pmatrix}$

   (c) $\begin{pmatrix} \frac{x}{2} & \frac{y}{3} \\ \frac{z}{4} & \frac{w}{7} \end{pmatrix} = \begin{pmatrix} \frac{-5}{2} & \frac{8}{6} \\ \frac{-12}{4} & 12 \end{pmatrix}$

6. Give the additive inverse of each of the following.

(a) $\begin{pmatrix} 7 & 6 \\ 3 & 4 \end{pmatrix}$

(c) $\begin{pmatrix} -\frac{1}{2} & \frac{1}{4} \\ 8 & -7 \end{pmatrix}$

(b) $\begin{pmatrix} -1.8 & 1.7 \\ 3.9 & -4.2 \end{pmatrix}$

(d) $\begin{pmatrix} 0 & 0 \\ -7.6 & 8.7 \end{pmatrix}$

7. Find the following sums.

(a) $\left[\begin{pmatrix} 3 & 4 \\ 5 & 6 \end{pmatrix} + \begin{pmatrix} 8 & 2 \\ 7 & 1 \end{pmatrix}\right] + \begin{pmatrix} 1 & 3 \\ -2 & 1 \end{pmatrix}$

(b) $\begin{pmatrix} 3 & 4 \\ 5 & 6 \end{pmatrix} + \left[\begin{pmatrix} 8 & 2 \\ 7 & 1 \end{pmatrix} + \begin{pmatrix} 1 & 3 \\ -2 & 1 \end{pmatrix}\right]$

(c) $\left[\begin{pmatrix} -1 & -6 \\ -4 & 3 \end{pmatrix} + \begin{pmatrix} -7 & 2 \\ -9 & -7 \end{pmatrix}\right] + \begin{pmatrix} 4 & 3 \\ 7 & 9 \end{pmatrix}$

(d) $\left[\begin{pmatrix} 2 & 4 \\ -1 & -1 \end{pmatrix} + \begin{pmatrix} 8 & 2 \\ -11 & -7 \end{pmatrix}\right] + \begin{pmatrix} 4 & 1 \\ 0 & 3 \end{pmatrix}$

8. Find values for x, y, z, and w that make the following true statements.

(a) $\begin{pmatrix} 2 & 3 \\ 7 & 6 \end{pmatrix} + \begin{pmatrix} x & y \\ z & w \end{pmatrix} = \begin{pmatrix} 6 & -2 \\ -4 & 9 \end{pmatrix}$

(b) $\begin{pmatrix} 2 & -7 \\ 3 & 4 \end{pmatrix} + \begin{pmatrix} z & y \\ z & w \end{pmatrix} = \begin{pmatrix} -4 & 15 \\ 6 & 9 \end{pmatrix}$

(c) $\begin{pmatrix} \frac{1}{2} & \frac{3}{4} \\ \frac{5}{6} & \frac{1}{10} \end{pmatrix} + \begin{pmatrix} x & y \\ z & w \end{pmatrix} = \begin{pmatrix} \frac{3}{4} & 1 \\ 2 & \frac{7}{10} \end{pmatrix}$

(d) $\begin{pmatrix} 1.7 & 1.8 \\ 3.4 & 6.1 \end{pmatrix} + \begin{pmatrix} x & y \\ z & w \end{pmatrix} = \begin{pmatrix} 2.9 & 3.1 \\ 8.2 & 7.5 \end{pmatrix}$

(e) $\begin{pmatrix} -\frac{1}{2} & -\frac{1}{4} \\ \frac{5}{8} & \frac{3}{16} \end{pmatrix} + \begin{pmatrix} x & y \\ z & w \end{pmatrix} = \begin{pmatrix} \frac{5}{8} & \frac{3}{4} \\ \frac{5}{2} & \frac{3}{8} \end{pmatrix}$

(f) $\begin{pmatrix} -1.11 & 3.04 \\ -4.06 & 7.11 \end{pmatrix} + \begin{pmatrix} x & y \\ z & w \end{pmatrix} = \begin{pmatrix} 1.11 & 3.84 \\ 9.07 & -4.06 \end{pmatrix}$

9. A manufacturer has three factories, one in the east, one in the midwest, and one in the west. Each factory produces dresses in junior sizes and misses sizes in two price ranges labeled 1 and 2.

The quantities produced in each factory are given in the following matrices:

	East (E)		West (W)		Midwest (M)	
	1	2	1	2	1	2
Junior	200	150	150	100	500	300
Misses	300	200	300	150	450	400

(a) Express the total production in a single matrix T.
(b) How many price 1 dresses are manufactured by the company?
(c) How many junior dresses are manufactured by the company?
(d) How many price 2 dresses are manufactured by the company?
(e) How many misses dresses are manufactured by the company?

10. A city has two junior colleges, one in the northern section of the city and one in the southern section. Each college has two types of programs, one a liberal-arts program and the other a vocational program. The students enrolled in each program are given in the following matrices:

	North (N)		South (S)	
	L.A.	Voc.	L.A.	Voc.
Freshmen	300	400	550	675
Sophomores	250	300	480	520

Express the total enrollment in a single matrix E.

11. An aircraft company has two plants in a certain city. It has salaried and hourly employees. The number of men and women employed is shown in the following matrices:

	Plant I		Plant II	
	Men	Women	Men	Women
Salaried	1200	875	650	300
Hourly	2400	1250	1525	850

Express the total number of employees in a single matrix S.

4. MULTIPLICATION OF MATRICES

Geometrically, a matrix may be used to describe a transformation of points in space. For example, if equations specifying the new location (x', y') of a point with coordinates (x, y) are given by

$$x' = ax + by$$
$$y' = cx + dy$$

then the matrix

$$A = \begin{pmatrix} a & b \\ c & d \end{pmatrix}$$

may be used to describe the transformation. If another transformation is given by

$$B = \begin{pmatrix} e & f \\ g & h \end{pmatrix}$$

then the result of applying transformation A on top of transformation B is given by

$$\begin{pmatrix} ae + bg & af + bh \\ ce + dg & cf + dh \end{pmatrix}$$

We shall not go into details of transformations here, but we mention that the foregoing gives rise to an operation with matrices called their **product**. If

$$A = \begin{pmatrix} a & b \\ c & d \end{pmatrix}$$

and

$$B = \begin{pmatrix} e & f \\ g & h \end{pmatrix}$$

are two matrices, then their **product** $A \cdot B$ is defined to be

$$A \cdot B = \begin{pmatrix} a & b \\ c & d \end{pmatrix} \cdot \begin{pmatrix} e & f \\ g & h \end{pmatrix} = \begin{pmatrix} ae + bg & af + bh \\ ce + dg & cf + dh \end{pmatrix}$$

The element in the first row and second column of the product $A \cdot B$ is the sum of elements each of which is the product of an element of the first row of A multiplied by the corresponding element from the second column of B. Thus

$$\begin{pmatrix} a & b \\ * & * \end{pmatrix} \cdot \begin{pmatrix} * & f \\ * & h \end{pmatrix} = \begin{pmatrix} * & af + bh \\ * & * \end{pmatrix}$$

If

$$A = \begin{pmatrix} 3 & 4 \\ 5 & 1 \end{pmatrix}$$

and

$$B = \begin{pmatrix} -1 & 2 \\ 3 & 1 \end{pmatrix}$$

then

$$A \cdot B = \begin{pmatrix} 3 & 4 \\ 5 & 1 \end{pmatrix} \cdot \begin{pmatrix} -1 & 2 \\ 3 & 1 \end{pmatrix}$$

$$= \begin{pmatrix} (3)(-1) + (4)(3) & (3)(2) + (4)(1) \\ (5)(-1) + (1)(3) & (5)(2) + (1)(1) \end{pmatrix}$$

$$= \begin{pmatrix} 9 & 10 \\ -2 & 11 \end{pmatrix}$$

However,

$$B \cdot A = \begin{pmatrix} -1 & 2 \\ 3 & 1 \end{pmatrix} \cdot \begin{pmatrix} 3 & 4 \\ 5 & 1 \end{pmatrix}$$

$$= \begin{pmatrix} (-1)(3) + (2)(5) & (-1)(4) + (2)(1) \\ (3)(3) + (1)(5) & (3)(4) + (1)(1) \end{pmatrix}$$

$$= \begin{pmatrix} 7 & -2 \\ 14 & 13 \end{pmatrix}$$

Notice that matrix multiplication is not necessarily commutative. $A \cdot B$ and $B \cdot A$ are not necessarily equal.

Let us consider another special 2×2 matrix in which the elements on the main diagonal are 1:

$$\begin{pmatrix} 1 & 0 \\ 0 & 1 \end{pmatrix}$$

Notice the product of this matrix and another 2×2 matrix:

$$\begin{pmatrix} 3 & 4 \\ 1 & 2 \end{pmatrix} \cdot \begin{pmatrix} 1 & 0 \\ 0 & 1 \end{pmatrix} = \begin{pmatrix} (3)(1) + (4)(0) & (3)(0) + (4)(1) \\ (1)(1) + (2)(0) & (1)(0) + (2)(1) \end{pmatrix}$$

$$= \begin{pmatrix} 3 & 4 \\ 1 & 2 \end{pmatrix}$$

In general,

$$\begin{pmatrix} a & b \\ c & d \end{pmatrix} \cdot \begin{pmatrix} 1 & 0 \\ 0 & 1 \end{pmatrix} = \begin{pmatrix} a \cdot 1 + b \cdot 0 & a \cdot 0 + b \cdot 1 \\ c \cdot 1 + d \cdot 0 & c \cdot 0 + d \cdot 1 \end{pmatrix} = \begin{pmatrix} a & b \\ c & d \end{pmatrix}$$

It is clear then why this matrix is called the multiplicative **identity matrix.**

Observe that the identity matrix serves the same function in matrix multiplication as 1 does in multiplication of real numbers.

Observe the following multiplications:

1. $\left[\begin{pmatrix} 3 & 1 \\ 2 & 1 \end{pmatrix} \cdot \begin{pmatrix} 1 & 0 \\ 2 & 1 \end{pmatrix}\right] \cdot \begin{pmatrix} 1 & 2 \\ 1 & 1 \end{pmatrix} = \begin{pmatrix} 5 & 1 \\ 4 & 1 \end{pmatrix} \cdot \begin{pmatrix} 1 & 2 \\ 1 & 1 \end{pmatrix} = \begin{pmatrix} 6 & 11 \\ 5 & 9 \end{pmatrix}$

$\begin{pmatrix} 3 & 1 \\ 2 & 1 \end{pmatrix} \cdot \left[\begin{pmatrix} 1 & 0 \\ 2 & 1 \end{pmatrix} \cdot \begin{pmatrix} 1 & 2 \\ 1 & 1 \end{pmatrix}\right] = \begin{pmatrix} 3 & 1 \\ 2 & 1 \end{pmatrix} \cdot \begin{pmatrix} 1 & 2 \\ 3 & 5 \end{pmatrix} = \begin{pmatrix} 6 & 11 \\ 5 & 9 \end{pmatrix}$

Hence

$\left[\begin{pmatrix} 3 & 1 \\ 2 & 1 \end{pmatrix} \cdot \begin{pmatrix} 1 & 0 \\ 2 & 1 \end{pmatrix}\right] \cdot \begin{pmatrix} 1 & 2 \\ 1 & 1 \end{pmatrix} = \begin{pmatrix} 3 & 1 \\ 2 & 1 \end{pmatrix} \cdot \left[\begin{pmatrix} 1 & 0 \\ 2 & 1 \end{pmatrix} \cdot \begin{pmatrix} 1 & 2 \\ 1 & 1 \end{pmatrix}\right]$

2. $\left[\begin{pmatrix} -1 & -1 \\ 0 & 1 \end{pmatrix} \cdot \begin{pmatrix} 2 & 1 \\ 1 & 2 \end{pmatrix}\right] \cdot \begin{pmatrix} -1 & 1 \\ 0 & 0 \end{pmatrix} = \begin{pmatrix} -3 & -3 \\ 1 & 2 \end{pmatrix} \cdot \begin{pmatrix} -1 & 1 \\ 0 & 0 \end{pmatrix} = \begin{pmatrix} 3 & -3 \\ -1 & 1 \end{pmatrix}$

$\begin{pmatrix} -1 & -1 \\ 0 & 1 \end{pmatrix} + \left[\begin{pmatrix} 2 & 1 \\ 1 & 2 \end{pmatrix} \cdot \begin{pmatrix} -1 & 1 \\ 0 & 0 \end{pmatrix}\right] = \begin{pmatrix} -1 & -1 \\ 0 & 1 \end{pmatrix} \cdot \begin{pmatrix} -2 & 2 \\ -1 & 1 \end{pmatrix} = \begin{pmatrix} 3 & -3 \\ -1 & 1 \end{pmatrix}$

Hence

$\left[\begin{pmatrix} -1 & -1 \\ 0 & 1 \end{pmatrix} \cdot \begin{pmatrix} 2 & 1 \\ 1 & 2 \end{pmatrix}\right] \cdot \begin{pmatrix} -1 & 1 \\ 0 & 0 \end{pmatrix} = \begin{pmatrix} -1 & -1 \\ 0 & 1 \end{pmatrix} \cdot \left[\begin{pmatrix} 2 & 1 \\ 1 & 2 \end{pmatrix} \cdot \begin{pmatrix} -1 & 1 \\ 0 & 0 \end{pmatrix}\right]$

In general

$\left[\begin{pmatrix} a & b \\ c & d \end{pmatrix} \cdot \begin{pmatrix} e & f \\ g & h \end{pmatrix}\right] \cdot \begin{pmatrix} w & x \\ y & z \end{pmatrix} = \begin{pmatrix} a & b \\ c & d \end{pmatrix} \cdot \left[\begin{pmatrix} e & f \\ g & h \end{pmatrix} \cdot \begin{pmatrix} w & x \\ y & z \end{pmatrix}\right]$

Because the order in which matrices are grouped in multiplication does not matter, we say that matrix multiplication is **associative**.

EXERCISE 3

1. Find the products $A \cdot B$ and $B \cdot A$.

(a) $A = \begin{pmatrix} 1 & 2 \\ 4 & 7 \end{pmatrix}$　　$B = \begin{pmatrix} 3 & 7 \\ 9 & 15 \end{pmatrix}$

(b) $A = \begin{pmatrix} 1 & 2 \\ 6 & 4 \end{pmatrix}$　　$B = \begin{pmatrix} 2 & 9 \\ -1 & 11 \end{pmatrix}$

2. Find the products $A \cdot B$ and $B \cdot A$.

(a) $A = \begin{pmatrix} 0 & 1 \\ 1 & 0 \end{pmatrix}$ $B = \begin{pmatrix} 3 & -1 \\ 4 & 7 \end{pmatrix}$

(b) $A = \begin{pmatrix} -3 & 4 \\ -2 & 1 \end{pmatrix}$ $B = \begin{pmatrix} 4 & -3 \\ 1 & -2 \end{pmatrix}$

3. Find the products $A \cdot B$ and $B \cdot A$.

(a) $A = \begin{pmatrix} -5 & 8 \\ 4 & 6 \end{pmatrix}$ $B = \begin{pmatrix} 1 & 7 \\ 6 & 0 \end{pmatrix}$

(b) $A = \begin{pmatrix} 81 & 76 \\ 14 & 12 \end{pmatrix}$ $B = \begin{pmatrix} 2 & 1 \\ -3 & 2 \end{pmatrix}$

4. Find the products $A \cdot B$ and $B \cdot A$.

$A = \begin{pmatrix} 1 & 0 \\ 1 & 1 \end{pmatrix}$ $B = \begin{pmatrix} -3 & -4 \\ 9 & -6 \end{pmatrix}$

$A = \begin{pmatrix} 2 & 4 \\ -1 & -2 \end{pmatrix}$ $B = \begin{pmatrix} 0 & 2 \\ 0 & -1 \end{pmatrix}$

5. Find the products $A \cdot B$ and $B \cdot A$.

$A = \begin{pmatrix} -1 & -3 \\ 7 & -6 \end{pmatrix}$ $B = \begin{pmatrix} -6 & 4 \\ 3 & 1 \end{pmatrix}$

$A = \begin{pmatrix} 20 & 30 \\ 40 & 50 \end{pmatrix}$ $B = \begin{pmatrix} -1 & -4 \\ -2 & -7 \end{pmatrix}$

6. Use the following matrices to find the products in (a) through (i) below.

$A = \begin{pmatrix} 1 & 3 \\ 2 & 1 \end{pmatrix}$ $B = \begin{pmatrix} -1 & 3 \\ -4 & -2 \end{pmatrix}$ $C = \begin{pmatrix} -2 & -3 \\ 1 & 4 \end{pmatrix}$

$D = \begin{pmatrix} 4 & 3 \\ -5 & 7 \end{pmatrix}$ $E = \begin{pmatrix} -3 & -1 \\ 0 & 1 \end{pmatrix}$ $F = \begin{pmatrix} 4 & 2 \\ -4 & -2 \end{pmatrix}$

(a) $(A \cdot B) \cdot C$ (d) $(A \cdot C) \cdot F$ (g) $(C \cdot F) \cdot D$
(b) $(D \cdot C) \cdot B$ (e) $(C \cdot E) \cdot F$ (h) $(B \cdot E) \cdot D$
(c) $E \cdot (F \cdot D)$ (f) $(D \cdot F) \cdot E$ (i) $(C \cdot A) \cdot D$

7. Given

$$A = \begin{pmatrix} 3 & 4 \\ 2 & 5 \end{pmatrix}$$

(a) Find $A^2 = A \cdot A$.
(b) Find $A^3 = A^2 \cdot A$.

5. MULTIPLICATIVE INVERSES OF 2 × 2 MATRICES

Every 2 × 2 matrix has an additive inverse. We now ask: Does every 2 × 2 matrix have a **multiplicative inverse**?

Let us try to find the multiplicative inverse of the matrix

$$\begin{pmatrix} 3 & 1 \\ 2 & 4 \end{pmatrix}$$

We are looking for the matrix

$$\begin{pmatrix} x & y \\ z & w \end{pmatrix}$$

such that

$$\begin{pmatrix} 3 & 1 \\ 2 & 4 \end{pmatrix} \cdot \begin{pmatrix} x & y \\ z & w \end{pmatrix} = \begin{pmatrix} 1 & 0 \\ 0 & 1 \end{pmatrix}$$

We see that

$$\begin{pmatrix} 3 & 1 \\ 2 & 4 \end{pmatrix} \cdot \begin{pmatrix} x & y \\ z & w \end{pmatrix} = \begin{pmatrix} 3x + z & 3y + w \\ 2x + 4z & 2y + 4w \end{pmatrix}$$

We wish this product to equal

$$\begin{pmatrix} 1 & 0 \\ 0 & 1 \end{pmatrix}$$

and therefore we have

$$\begin{pmatrix} 3x + z & 3y + w \\ 2x + 4z & 2y + 4w \end{pmatrix} = \begin{pmatrix} 1 & 0 \\ 0 & 1 \end{pmatrix}$$

and hence

(1) $3x + z = 1$
(2) $3y + w = 0$
(3) $2x + 4z = 0$
(4) $2y + 4w = 1$

Multiplying equation (2) by 4 and subtracting equation (4) from it, we have

$$\begin{aligned} 12y + 4w &= 0 \\ \underline{2y + 4w} &= 1 \\ 10y &= -1 \\ y &= -\tfrac{1}{10} \end{aligned}$$

Substituting $y = -\frac{1}{10}$ in equation (2) we have

$$-\frac{3}{10} + w = 0$$
$$w = \frac{3}{10}$$

Multiplying equation (1) by 4 and subtracting equation (3) from it we have

$$12x + 4z = 4$$
$$\underline{2x + 4z = 0}$$
$$10x \qquad = 4$$
$$x \qquad = \frac{2}{5}$$

Substituting $x = \frac{2}{5}$ in equation (1) we have

$$\frac{6}{5} + z = 1$$
$$z = -\frac{1}{5}$$

The values $\frac{2}{5}$, $-\frac{1}{10}$, $-\frac{1}{5}$, and $\frac{3}{10}$, for x, y, z, and w yield the matrix

$$\begin{pmatrix} \frac{2}{5} & -\frac{1}{10} \\ -\frac{1}{5} & \frac{3}{10} \end{pmatrix}$$

Now we must check to see whether this matrix is the multiplicative inverse of the given matrix. We do this by checking the multiplication:

$$\begin{pmatrix} 3 & 1 \\ 2 & 4 \end{pmatrix} \cdot \begin{pmatrix} \frac{2}{5} & -\frac{1}{10} \\ -\frac{1}{5} & \frac{3}{10} \end{pmatrix} = \begin{pmatrix} \frac{6}{5} - \frac{1}{5} & -\frac{3}{10} + \frac{3}{10} \\ \frac{4}{5} - \frac{4}{5} & -\frac{2}{10} + \frac{12}{10} \end{pmatrix}$$

$$= \begin{pmatrix} \frac{5}{5} & 0 \\ 0 & \frac{10}{10} \end{pmatrix} = \begin{pmatrix} 1 & 0 \\ 0 & 1 \end{pmatrix}$$

The fact that *one* 2 × 2 matrix has an inverse does not mean that *every* 2 × 2 matrix has an inverse. Consider the matrix

$$\begin{pmatrix} 2 & 1 \\ 4 & 2 \end{pmatrix}$$

Can we find a matrix

$$\begin{pmatrix} x & y \\ z & w \end{pmatrix}$$

such that

$$\begin{pmatrix} 2 & 1 \\ 4 & 2 \end{pmatrix} \cdot \begin{pmatrix} x & y \\ z & w \end{pmatrix} = \begin{pmatrix} 1 & 0 \\ 0 & 1 \end{pmatrix}$$

Assuming that we can find such a matrix, we use the same method as previously used and we find that we must solve the following equations:

(1) $2x + z = 1$
(2) $2y + w = 0$
(3) $4x + 2z = 0$
(4) $4y + 2w = 1$

Multiplying equation (1) by 2 and subtracting it from equation (3) we have the impossible result $0 = 2$.

This tells us that our assumption that the given matrix has an inverse is false. Hence we cannot say that every 2×2 matrix has a multiplicative inverse.

It can be proved that every 2×2 matrix

$$\begin{pmatrix} a & b \\ c & d \end{pmatrix}$$

has a multiplicative inverse provided $ad - bc \neq 0$. If $ad - bc = 0$, then the matrix does not have a multiplicative inverse.

EXERCISE 4

1. Find the multiplicative inverse of each of the following.

(a) $\begin{pmatrix} 2 & 2 \\ 2 & 4 \end{pmatrix}$ (b) $\begin{pmatrix} 3 & 1 \\ 1 & 2 \end{pmatrix}$

2. Find the multiplicative inverse of each of the following.

(a) $\begin{pmatrix} 1 & 2 \\ 1 & 1 \end{pmatrix}$ (b) $\begin{pmatrix} 2 & 0 \\ 3 & 4 \end{pmatrix}$

3. Find the multiplicative inverse of each of the following.

(a) $\begin{pmatrix} 0 & 1 \\ 2 & 1 \end{pmatrix}$ (b) $\begin{pmatrix} 3 & -2 \\ -1 & 2 \end{pmatrix}$

4. Show that the following matrices do not have a multiplicative inverse.

(a) $\begin{pmatrix} 8 & 1 \\ 16 & 2 \end{pmatrix}$ (b) $\begin{pmatrix} 2 & 5 \\ 4 & 10 \end{pmatrix}$ (c) $\begin{pmatrix} 3 & 2 \\ 6 & 4 \end{pmatrix}$

5. Show that the following matrices do not have a multiplicative inverse.

(a) $\begin{pmatrix} 9 & 6 \\ 3 & 2 \end{pmatrix}$ (b) $\begin{pmatrix} 2 & 4 \\ 4 & 8 \end{pmatrix}$ (c) $\begin{pmatrix} 3 & 6 \\ 4 & 8 \end{pmatrix}$

6. Which of the following matrices have multiplicative inverses? Which do not?

(a) $\begin{pmatrix} 3 & 4 \\ 1 & -2 \end{pmatrix}$ (c) $\begin{pmatrix} 4 & 8 \\ 8 & 16 \end{pmatrix}$

(b) $\begin{pmatrix} 6 & -4 \\ -12 & 8 \end{pmatrix}$ (d) $\begin{pmatrix} 3 & -2 \\ 1 & -7 \end{pmatrix}$

7. Which of the following matrices have multiplicative inverses? Which do not?

(a) $\begin{pmatrix} 7 & 0 \\ 0 & 1 \end{pmatrix}$ (c) $\begin{pmatrix} 5 & 6 \\ 10 & 12 \end{pmatrix}$

(b) $\begin{pmatrix} -9 & -6 \\ -6 & -4 \end{pmatrix}$ (d) $\begin{pmatrix} -1 & 3 \\ -2 & 7 \end{pmatrix}$

GOTTFRIED WILHELM VON LEIBNITZ (1646–1716), a genius in many fields including mathematics, advocated the binary system of notation and he attached mystic significance to this system, believing it was the "image of Creation." He imagined that unity (one) represented God and zero the void from which the Supreme Being drew all things, just as one and zero are the only symbols needed to express all numbers in the binary system. In 1672 Leibnitz invented a calculating machine that could add, subtract, multiply, and divide. He made only a few of these machines, one of which, still in working condition, was in the Leibnitz museum in Hanover prior to World War II.

Computers and Numeration Systems/10

1. A BRIEF HISTORY OF THE DIGITAL COMPUTER

Mathematics is fascinating, but computation is boring. Because of this, men have always tried to devise computing tools to make this drudgery easier. The earliest calculating device (other than the fingers) was the abacus. This was followed by Napier's Bones (1614), Pascal's Calculator (1642), Leibnitz's Reckoning Machine (1671), and Babbage's Difference Machine (1830). The latest of these computing devices is the **digital computer**.

In 1642 a French mathematician, Blaise Pascal, invented what was probably the first mechanical adding machine. This modest machine, with its hand-fashioned rachets and gears, began the chain of events leading to today's digital computer.

Early in the nineteenth century, Charles Babbage, a young Englishman, began work on a machine that was to compute mathematical tables. His "difference machine" needed a human operator to start the calculation and it then proceeded automatically until all the answers had been printed. Although Babbage's machine was never fully successful, many of the ideas that he proposed were adopted approximately one hundred years later.

Around 1890 Herman Hollerith, who worked for the United States Bureau of the Census, devised a system in which holes punched in cards represented numbers, letters, and symbols. His next project was the development of an electromechanical device that "read" the cards and tabulated the data they contained. Such punched cards are seen every day in pay checks, registration cards, bills, and the like.

209

At the turn of the century desk calculators had become compact and reliable, yet their new efficiency was not enough. Need was fast arising for a machine that would perform a sequence of operations automatically while saving intermediate results for later use. The first attempts at this type of device was an enlargement on the desk calculator. This resulted in a maze of wheels, shafts, gears, and the like, all requiring precision adjustment and intricate balance. This soon became impractical both in size and cost. The next step was to enlarge the capabilities of the punched-card machines. Even though these machines could do only one step at a time, it was easier and faster to pass decks of cards through many times than to do the calculations on a desk calculator.

Around 1937 George Slibitz and Howard Aiken began to work independently on a sequentially operated digital computer. The first digital computer was unveiled in 1940 by George Slibitz and the Bell Telephone Laboratories at the American Mathematical Society meeting in Hanover, New Hampshire. Howard Aiken, in collaboration with International Business Machines (IBM), designed the Mark I computer which was announced to the public in 1944. It was installed at Harvard University.

World War II was one of the major factors that hastened the development of the digital computer. The need for more sophisticated weapons demanded more extensive analyses. The computer filled this demand. More and more research was concentrated on improving its capabilities and speed. Soon after the end of the war, Remington Rand Corporation introduced the Eniac computer. In the Eniac the electrical relays, used in older computers to represent the digits, had been replaced by electron tubes. This replacement increased the speed of computation by computers as much as 100 per cent.

One of the most important inventions that furthered the development of computers was a tiny ferromagnetic ring to replace the electron tube for digit storage. These rings, or **cores** as they are sometimes called, are very reliable, generate no heat, require little current to operate, and permit a great reduction in computer storage size.

The computer era has just begun, and it is here to stay. The authors of this book hope that the brief material presented here will give the reader a better understanding of these machines. The basic fact we must remember is that computers do not think. They are capable only of performing the manipulations wired into their systems and executing the sequence of instructions supplied by men and women.

2. COMPUTER COMPONENTS

The operation of a digital computer can be divided roughly into five phases: (1) **input**, (2) **control**, (3) **storage** and **memory**, (4) **processing**, and (5) **output.** We shall outline these phases, omitting the technical details.

INPUT

Input is the **information**, which may be in the form of numbers, letters, or symbols, submitted to the machine. The input may be on punched cards, paper or magnetic tape, or entered manually by a keyboard or typewriter.

CONTROL

Computers operate under direction of a **control** unit. This unit interprets and executes the sequence of instructions called a **program**. In today's computers the program is stored in the machine's memory.

STORAGE AND MEMORY

Data is retained internally in what is called the **memory** or **core**. The memory is a multitude of tiny ferromagnetic rings arranged in groups. These groups, often called **words**, may contain a number or a program instruction. Each word, regardless of the content, can be located by the control unit.

Other storage devices are magnetic tape, magnetic disks, and magnetic drums. All three storage devices use patterns of magnetic spots to represent data. Magnetic-tape units are similar to ordinary tape recorders using heads to record or "play back" data. Magnetic-disk storage may be thought of as a stack of phonograph records. Data stored on either side of the disks may be located by sensing arms that move in and out as the disks rotate. A magnetic drum is a cylinder that revolves several thousand times per second past a series of read-write heads.

PROCESSING

The **processing unit** performs the arithmetical operations of addition, subtraction, multiplication, and division. All complex mathematical problems must be broken down into combinations of these operations. The processing unit is also capable of logical decisions when comparing two words (greater than, less than, or equal to). It can also distinguish positive, negative, and zero values.

OUTPUT

Output is the answer to the problem. This may be obtained in printed form or on punched cards or magnetic tape, depending on the program.

3. HOW FAST ARE COMPUTERS?

Timing of machine operations is based on access time, called a **machine cycle**. A **cycle** is the amount of time required to transmit information between the storage area and the arithmetic unit. All instructions are timed in multiples of cycles—for example, an instruction to add may take four cycles to perform. The cycle time for the IBM 704, introduced in 1956, is 12 microseconds (0.000012 second). The IBM 7094, released in 1961, has cycle time of 2 microseconds (0.000002 second). Addition and subtraction of integers using the 7094 takes two cycles. Multiplication and division of integers on the 7094 take two to five cycles and three to eight cycles respectively, depending on the magnitude of the numbers. Non-integral numbers require more time to manipulate.

The most recent computers, the so-called "third generation," are even faster. IBM's System/360-65 has a cycle time of 0.75 microsecond, and the CDC* 6800 can add two decimal numbers in 0.0000001 second.

The fact that the 6800 can add 10,000,000 decimal numbers in one second emphasizes the speed of computers. The machines of tomorrow certainly will surpass present-day standards and no one can predict how fast they will be in the future.

4. THE BINARY NUMERATION SYSTEM

As far as we know, the core storage units of all of today's large computers use tiny ferromagnetic rings to retain data. These rings are about the size of the printed letter "o." The rings are threaded on wire in such a manner that when a current is passed through the wire, a magnetic field is set up in the ring. Passing the current in one direction sets up a "+" field representing a "1" or "yes" condition; changing the direction

---

* Control Data Corporation.

of the current changes the polarity of the ring to a "$-$" field representing a "0" or a "no" condition. The representation of a "**bit**" (a contraction of binary digit) of data by either a 1 or a 0 is the basis of the **binary numeration system**. As we shall see later, many binary numerals are required to represent equivalent decimal numerals and thus the core storage device must contain many magnetic rings. (The IBM 7094 has 32,768 words; each word is represented by 36 magnetic rings.)

The binary system of notation is easier to understand if we relate it to the familiar Hindu-Arabic system of numeration. The **Hindu-Arabic** system of numeration is a **place-value** system that uses ten symbols, 0, 1, 2, 3, 4, 5, 6, 7, 8, and 9, to represent the numbers zero, one, two, three, four, five, six, seven, eight, and nine. The next number, the compounding point, is ten (represented by 10) and names the **base** of the system. Because the base of this system of numeration is ten, it is called the **decimal system** (from the Latin word "decem" meaning ten).

The symbols 0, 1, 2, 3, 4, 5, 6, 7, 8, and 9 are called **digits**. Any number greater than nine can be represented as a combination of these digits. The **value** of a number represented by a combination of digits is the sum of products determined by our system of place value. The place value of a digit is determined by its location within the numeral, not by the digit itself. Table 10.1 shows the place value of various positions in the decimal system.

TABLE 10.1 *Place-value chart for base ten*

0	1	1	10^0	one
1	10	10 × 1	10^1	ten
2	100	10 × 10	10^2	one hundred
3	1,000	10 × 100	10^3	one thousand
4	10,000	10 × 1000	10^4	ten thousand
5	100,000	10 × 10,000	10^5	hundred thousand
⋮	⋮	⋮	⋮	

If we write a decimal numeral in **expanded notation**, the idea of place value becomes clearer. The number one thousand six hundred twenty-four may be written as:

$$\overset{①②③④}{1624} = (\text{one} \times \text{one thousand}) + (\text{six} \times \text{one hundred}) + (\text{two} \times \text{ten}) + (\text{four} \times \text{one})$$

The small numerals within the circles above the digits of the numeral indicate the position of the digits. We may also write

$$1624 = (1 \times 1000) + (6 \times 100) + (2 \times 10) + (4 \times 1)$$

or, using powers of the base,

$$1624 = (1 \times 10^3) + (6 \times 10^2) + (2 \times 10^1) + (4 \times 10^0)$$

Notice that the exponents of 10 correspond to the numerals in the circles indicating the positions of the various digits.

The value of a number is just the sum of products. Thus

$$1624 = (1 \times 1000) + (6 \times 100) + (2 \times 10) + (4 \times 1)$$
$$= 1000 + 600 + 20 + 4$$

If we extend this scheme to the right of the units place, naming the positions $-1, -2, -3, \ldots$, our place-value chart can be extended by adding the places indicated in Table 10.2.

TABLE 10.2 *Place-value chart for base ten*

$\vdots$	$\vdots$	$\vdots$	$\vdots$	$\vdots$	$\vdots$
-5	.00001	$\dfrac{1}{100,000}$	$\dfrac{1}{10 \times 10,000}$	10^{-5}	one hundred-thousandth
-4	.0001	$\dfrac{1}{10,000}$	$\dfrac{1}{10 \times 1000}$	10^{-4}	one ten-thousandth
-3	.001	$\dfrac{1}{1000}$	$\dfrac{1}{10 \times 100}$	10^{-3}	one-thousandth
-2	.01	$\dfrac{1}{100}$	$\dfrac{1}{10 \times 10}$	10^{-2}	one-hundredth
-1	.1	$\dfrac{1}{10}$	$\dfrac{1}{10 \times 1}$	10^{-1}	one-tenth

Writing the decimal fraction 0.839 in expanded notation we have

$$0.839 = (8 \times \text{one-tenth}) + (3 \times \text{one-hundredth})$$
$$+ (9 \times \text{one-thousandth})$$

$$= (8 \times .1) + (3 \times .01) + (9 \times .001)$$

$$= \left(8 \times \frac{1}{10}\right) + \left(3 \times \frac{1}{10^2}\right) + \left(9 \times \frac{1}{10^3}\right)$$

$$= (8 \times 10^{-1}) + (3 \times 10^{-2}) + (9 \times 10^{-3})$$

Again observe that the exponents of 10 match the position numerals in the small circles, and that the value of the number is the sum of products.

The **binary system** of numeration (base two) is also a place-value system with two digits, 0 and 1, which represent the numbers zero and one, respectively. All numbers, symbols, and alphabetic characters used internally by digital computers must be combinations of these two symbols.

In base ten, the place values are powers of ten. In base two, the place values are powers of two:

$$
\begin{aligned}
&\vdots \\
\text{two}^0 &= \text{one} \\
\text{two}^1 &= \text{two} \\
\text{two}^2 &= \text{four} \\
\text{two}^3 &= \text{eight} \\
&\vdots
\end{aligned}
$$

The binary numeral for two is $10 = (1 \times \text{two}^1) + (0 \times \text{two}^0)$. The first few binary numerals and their decimal equivalents are:

Binary Notation	Decimal Notation
0	0
1	1
10 = 1 two + 0 ones	2
11 = 1 two + 1 one	3
100 = 1 four + 0 twos + 0 ones	4
⋮	⋮
1000 = 1 eight + 0 fours + 0 twos + 0 ones	8
1001 = 1 eight + 0 fours + 0 twos + 1 one	9
⋮	⋮

Table 10.3 is the place-value chart for the binary system.

Now let us examine some binary numerals and find their decimal equivalents. Notice that in the first expansion we spell out the "two." We

TABLE 10.3 *Place-value chart for the binary system of notation*

Binary			Decimal Equivalent			
⋮	⋮	⋮	⋮	⋮	⋮	⋮
−5	two^{-5}	.00001	2^{-5}	$\frac{1}{2^5}$	$\frac{1}{32}$	.03125
−4	two^{-4}	.0001	2^{-4}	$\frac{1}{2^4}$	$\frac{1}{16}$	.0625
−3	two^{-3}	.001	2^{-3}	$\frac{1}{2^3}$	$\frac{1}{8}$	.125
−2	two^{-2}	.01	2^{-2}	$\frac{1}{2^2}$	$\frac{1}{4}$	.25
−1	two^{-1}	.1	2^{-1}	$\frac{1}{2^1}$	$\frac{1}{2}$	.5
0	two^{0}	1.	2^{0}	$\frac{1}{2^0}$	$\frac{1}{1}$	1.
1	two^{1}	10.	2^{1}	$\frac{2^1}{1}$	$\frac{2}{1}$	2.
2	two^{2}	100.	2^{2}	$\frac{2^2}{1}$	$\frac{4}{1}$	4.
3	two^{3}	1000.	2^{3}	$\frac{2^3}{1}$	$\frac{8}{1}$	8.
4	two^{4}	10000.	2^{4}	$\frac{2^4}{1}$	$\frac{16}{1}$	16.
5	two^{5}	100000.	2^{5}	$\frac{2^5}{1}$	$\frac{32}{1}$	32.
⋮	⋮	⋮	⋮	⋮	⋮	⋮

do this because the binary system has no numeral "2." The second, third, and fourth lines in each example are decimal equivalents.

$$1011.1 = (1 \times \text{two}^3) + (0 \times \text{two}^2) + (1 \times \text{two}^1) + (1 \times \text{two}^0)$$
$$+ (1 \times \text{two}^{-1})$$
$$= (1 \times 2^3) + (0 \times 2^2) + (1 \times 2^1) + (1 \times 2^0) + (1 \times 2^{-1})$$
$$= 8 + 0 + 2 + 1 + 0.5$$
$$= 11.5$$

$$11001.01. = (1 \times two^4) + (1 \times two^3) + (0 \times two^2) + (0 \times two^1)$$
$$+ (1 \times two^0) + (0 \times two^{-1}) + (1 \times two^{-2})$$
$$= (1 \times 2^4) + (1 \times 2^3) + (0 \times 2^2) + (0 \times 2^1) + (1 \times 2^0)$$
$$+ (0 \times 2^{-1}) + (1 \times 2^{-2})$$
$$= 16 + 8 + 0 + 0 + 1 + 0 + 0.25$$
$$= 25.25$$

$$11.001 = (1 \times two^1) + (1 \times two^0) + (0 \times two^{-1}) + (0 \times two^{-2})$$
$$= (1 \times two^{-3})$$
$$= (1 \times 2^1) + (1 \times 2^0) + (0 \times 2^{-1}) + (0 \times 2^{-2}) + (1 \times 2^{-3})$$
$$= 2 + 1 + 0 + 0 + 0.125$$
$$= 3.125$$

If we wish to find a binary numeral equivalent to a decimal numeral, we use place value. Suppose we wish to find a binary numeral equivalent to the decimal numeral 85. We ask ourselves: What is the largest power of 2 less than 85? We see that the largest power of 2 less than 85 is $2^6 = 64$. Since we have one set of $2^6 = 64$, we put a "1" in position 6.

6	5	4	3	2	1	0
1						

To find the digit in position 5 we determine whether we can take a set of $2^5 = 32$ from $85 - 64 = 21$. Since we cannot take a set of 32 from 21, we place a "0" in position 5.

6	5	4	3	2	1	0
1	0					

To find the digit in position 4 we determine whether we can take a set of $2^4 = 16$ from 21. Since we can take a set of $2^4 = 16$ from 21, we put a "1" in position 4.

6	5	4	3	2	1	0
1	0	1				

Now we ask whether we can take a set of $2^3 = 8$ from $21 - 16 = 5$? Since we cannot take a set of $2^3 = 8$ from 5, we put a "0" in position 3.

6	5	4	3	2	1	0
1	0	1	0			

Can we take a set of $2^2 = 4$ from 5? Since we can take a set of $2^2 = 4$ from 5, we put a "1" in position 2.

6	5	4	3	2	1	0
1	0	1	0	1		

We see that we cannot take a set of $2^1 = 2$ from $5 - 4 = 1$; so we put a "0" in position 1.

6	5	4	3	2	1	0
1	0	1	0	1	0	

We can take a set of $2^0 = 1$ from 1, hence we put a "1" in the 0 position.

6	5	4	3	2	1	0
1	0	1	0	1	0	1

The binary equivalent of 85 is

$$1010101$$

The foregoing discussion is shown below in a shortened form.

$$85 = 64 + 0 + 16 + 0 + 4 + 0 + 1$$
$$= (1 \times 64) + (0 \times 32) + (1 \times 16) + (0 \times 8) + (1 \times 4) + (0 \times 2)$$
$$+ (1 \times 1)$$
$$= (1 \times two^6) + (0 \times two^5) + (1 \times two^4) + (0 \times two^3) + (1 \times two^2)$$
$$+ (0 \times two^1) + (1 \times two^0)$$
$$= 1,010,101$$

Similarly

$369 = 256 + 0 + 64 + 32 + 16 + 0 + 0 + 0 + 1$
$= (1 \times 256) + (0 \times 128) + (1 \times 64) + (1 \times 32) + (1 \times 16)$
$+ (0 \times 8) + (0 \times 4) + (0 \times 2) + (1 \times 1)$
$= (1 \times two^8) + (0 \times two^7) + (1 \times two^6) + (1 \times two^5)$
$+ (1 \times two^4) + (0 \times two^3) + (0 \times two^2)$
$+ (0 \times two^1) + (1 \times two^0)$
$= 101,110,001$

$77.5 - 64 + 0 + 0 + 8 + 4 + 0 + 1 + .5$
$= (1 \times 64) + (0 \times 32) + (0 \times 16) + (1 \times 8) + (1 \times 4) + (0 \times 2)$
$+ (1 \times 1) + (1 \times \frac{1}{2})$
$= (1 \times 64) + (0 \times 32) + (0 \times 16) + (1 \times 8) + (1 \times 4) + (0 \times 2)$
$+ (1 \times 1) + (1 \times 2^{-1})$
$= (1 \times two^6) + (0 \times two^5) + (0 \times two^4) + (1 \times two^3)$
$+ (1 \times two^2) + (0 \times two^1) + (1 \times two^0) + (1 \times two^{-1})$
$= 1,001,101.1$

Another method of converting whole numbers from decimal to binary notation involves repeated divisions. Each subsequent division is the subtraction of the next higher power of the base and the remainder is the digit for that particular place:

$$
\begin{array}{r|l|l}
2\underline{)33} & & \\
2\underline{)16} & R = 1 & 1 \times 2^0 \\
2\underline{)8} & R = 0 & 0 \times 2^1 \\
2\underline{)4} & R = 0 & 0 \times 2^2 \\
2\underline{)2} & R = 0 & 0 \times 2^3 \\
2\underline{)1} & R = 0 & 0 \times 2^4 \\
0 & R = 1 & 1 \times 2^5 \\
\end{array}
$$

To find the answer we read up; thus

$$33 \text{ (base ten)} = 100,001 \text{ (base two)}$$

Addition in the binary system is simple because only four addition facts need be learned. These are

$$0 + 0 = 0$$
$$1 + 0 = 1$$
$$0 + 1 = 1$$
$$1 + 1 = 10$$

These addition facts may be shown more compactly in an **addition table**. The addition table for the binary system is shown in Table 10.4.

TABLE 10.4

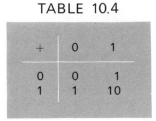

+	0	1
0	0	1
1	1	10

Just as a sum of ten or more in decimal addition results in a **regrouping** (usually called **carrying**) into the next higher position, a sum of two or more in binary addition results in a regrouping (called a **carry**) into the next higher position.

Study the following binary addition example. The explanation appears on the right; the carries are encircled.

$$\overset{①①\ ①①}{\begin{array}{r} 1\ 1\ ,0\ 1\ 1 \\ 1\ ,1\ 0\ 1 \\ \hline 1\ 0\ 1\ ,0\ 0\ 0 \end{array}}$$

$1 + 1 = 10$	Write the 0 and carry 1 (two^1).
$1 + 1 + 0 = 10$	Write the 0 and carry 1 (two^2).
$1 + 0 + 1 = 10$	Write the 0 and carry 1 (two^3).
$1 + 1 + 1 = 11$	Write the 1 and carry 1 (two^4).
$1 + 1 = 10$	Write the 10.

Other examples of binary addition include:

11,101	101	100	110
1,101	111	111	10
101,010	1,100	1,011	1,000

Subtraction in the binary system is done the same way as subtraction in the decimal system. Study the following examples.

Example 1.

$$\begin{array}{r} \overset{0\ \ 10}{1\ \ \cancel{1}\ \ 0} \\ 1 \\ \hline 1\ \ 0\ \ 1 \end{array}$$

One two in *this* place becomes two ones in *this* place.

Notice that 1 in any place becomes 10 in the next place to the right.

Example 2.

Step 1

```
            0 10
   1 0 1 1 0
       1 1
   _____
           1
```

Step 2

```
            10
         0  0 10
   1 0 1 1 0
        1 1
   _____
   1 0 0 1 1
```

Example 3.

Step 1

```
   0 10
   1 0 0 0 1
     1 1 1 1
   _____
           0
```

Step 2

```
     1 10
     10
   1 0 0 0 1
     1 1 1 1
   _____
           0
```

Step 3

```
       1 10
     1 10
     10
   1 0 0 0 1
     1 1 1 1
   _____
   0 0 0 1 0
```

Example 4

Step 1

```
   10
   1 0 0 0
     1 1 0
   _____
         0
```

Step 2

```
     1 10
     10
   1 0 0 0
     1 1 0
   _____
   0 0 1 0
```

EXERCISE 1

1. Write the following decimal numerals in expanded notation.
   - (a) 78
   - (b) 163
   - (c) 827
   - (d) 7619
   - (e) 82,000
   - (f) 130,000

2. Write the following decimal numerals in expanded notation.
   - (a) 93,000,000
   - (b) 18.2
   - (c) 132.01
   - (d) 1,671.321
   - (e) 876.007
   - (f) 96.1041

3. Convert the following binary numerals to equivalent decimal numerals.
   (a) 100   (c) 100001   (e) 101011111
   (b) 1000   (d) 10101011   (f) 10000000000

4. Convert the following binary numerals to equivalent decimal numerals.
   (a) 1011   (c) 11.11   (e) 1011.101
   (b) 11010   (d) 1.0011   (f) 1011100.00001

5. Convert the following decimal numerals to equivalent binary numerals.
   (a) 8   (c) 36   (e) 77
   (b) 13   (d) 69   (f) 98

6. Convert the following decimal numerals to equivalent binary numerals.
   (a) 112   (c) 3000   (e) 3012
   (b) 1024   (d) 1144   (f) 5149

7. Add. All numerals are in binary notation.
   (a) 100   (c) 1001
       11   101

   (b) 1001   (d) 1101
       1001   1001

8. Add. All numerals are in binary notation.
   (a) 1010.11   (c) 1011.1
       10.01   1.1

   (b) 1110.11   (d) 111.111
       100.11   101.101

9. Subtract. All numerals are in binary notation.
   (a) 1101   (b) 1001   (c) 1110
       101   1000   101

10. Subtract. All numerals are in binary notation.
    (a) 1011   (b) 11101   (c) 10001
        101   1010   110

5. OCTAL NUMERATION SYSTEM

Numerals written in the **octal** (base eight) system are represented by a combination of eight symbols, 0, 1, 2, 3, 4, 5, 6, 7. The octal system is also a place-value system. A place-value chart for base eight is given in Table 10.5.

TABLE 10.5 *Place-value chart for base eight*

Base Eight			Decimal Equivalent			
⋮	⋮	⋮	⋮	⋮	⋮	⋮
-4	eight^{-4}	.0001	8^{-4}	$\dfrac{1}{8^4}$	$\dfrac{1}{4096}$	$\doteq$ .000244
-3	eight^{-3}	.001	8^{-3}	$\dfrac{1}{8^3}$	$\dfrac{1}{512}$	$\doteq$ .001953
-2	eight^{-2}	.01	8^{-2}	$\dfrac{1}{8^2}$	$\dfrac{1}{64}$	.015625
-1	eight^{-1}	.1	8^{-1}	$\dfrac{1}{8^1}$	$\dfrac{1}{8}$	.125
0	eight^{0}	1.	8^{0}	$\dfrac{1}{8^0}$	$\dfrac{1}{1}$	1.
1	eight^{1}	10.	8^{1}	$\dfrac{8^1}{1}$	$\dfrac{8}{1}$	8.
2	eight^{2}	100.	8^{2}	$\dfrac{8^2}{1}$	$\dfrac{64}{1}$	64.
3	eight^{3}	1000.	8^{3}	$\dfrac{8^3}{1}$	$\dfrac{512}{1}$	512.
4	eight^{4}	10000.	8^{4}	$\dfrac{8^4}{1}$	$\dfrac{4096}{1}$	4096.
⋮	⋮	⋮	⋮	⋮	⋮	⋮

When we write **octal numerals** we think of objects grouped in sets of eight. Just as in base ten the representation of ten or more objects results in a regrouping or "carry" into the next higher position, so does the representation of eight or more objects result in a "carry" in base eight. In the following examples, the first lines represent the positional values in base eight; lines two, three, and four are the decimal equivalent. We use the notation $73_{(\text{eight})}$ to signify that the numeral 73 is written in base-eight notation.

Example 1.

$$73_{(eight)} = (7 \times eight^1) + (3 \times eight^0)$$
$$= (7 \times 8^1) + (3 \times 8^0)$$
$$= 56 + 3$$
$$= 59$$

Example 2.

$$165_{(eight)} = (1 \times eight^2) + (6 \times eight^1) + (5 \times eight^0)$$
$$= (1 \times 8^2) + (6 \times 8^1) + (5 \times 8^0)$$
$$= 64 + 48 + 5$$
$$= 117$$

Example 3.

$$427_{(eight)} = (4 \times eight^2) + (2 \times eight^1) + (7 \times eight^0)$$
$$= (4 \times 8^2) + (2 \times 8^1) + (7 \times 8^0)$$
$$= 256 + 16 + 7$$
$$= 279$$

From the foregoing examples we can see that the value of a number written in octal notation is the sum of the products formed by each digit multiplied by its place value.

To change a decimal numeral to an equivalent octal numeral, we must think of the number of objects grouped in set of $eight^0$, $eight^1$, $eight^2$, and so forth instead of sets of ten^0, ten^1, ten^2, and so forth. Suppose we wish to find the base-eight numeral equivalent to the decimal numeral 123.

We ask ourselves what is the largest power of 8 less than 123. Since $8^3 = 512$ we cannot take a set of 8^3 from 123; $8^2 = 64$ is less than 123 so to find the digit in position 2, we determine the greatest multiple of 64 less than 123.

$$
\begin{array}{r}
1 \\
64\overline{)123} \\
64 \\
\hline
59
\end{array}
$$

We see that we can subtract one $8^2 = 64$ from 123, hence we write "1" in position 2.

2	1	0
1		

To find the digit for position 1 we determine the largest multiple of $8^1 = 8$ less than $123 - 64 = 59$.

$$\begin{array}{r} 7 \\ 8\overline{)59} \\ 56 \\ \hline 3 \end{array}$$

We put a "7" in position 1.

2	1	0
1	7	

Since there is a remainder of 3 when we subtract $8 \times 7 = 56$ from 59 we put a "3" in position 0.

2	1	0
1	7	3

Hence

$$123_{(ten)} = 173_{(eight)}$$

The foregoing discussion is shown below in shortened form.

$$\begin{aligned} 123 &= 64 + 56 + 3 \\ &= (1 \times 64) + (7 \times 8) + (3 \times 1) \\ &= (1 \times eight^2) + (7 \times eight^1) + (3 \times eight^0) \\ &= 173_{(eight)} \end{aligned}$$

Similarly,

$$\begin{aligned} 1096 &= 1024 + 64 + 8 \\ &= (2 \times 512) + (1 \times 64) + (1 \times 8) + (0 \times 1) \\ &= (2 \times eight^3) + (1 \times eight^2) + (1 \times eight^1) + (0 \times eight^0) \\ &= 2110_{(eight)} \end{aligned}$$

$$\begin{aligned} 975 &= 512 + 448 + 8 + 7 \\ &= (1 \times 512) + (7 \times 64) + (1 \times 8) + (7 \times 1) \\ &= (1 \times eight^3) + (7 \times eight^2) + (1 \times eight^1) + (7 \times eight^0) \\ &= 1717 \end{aligned}$$

A base-ten numeral also may be converted to an equivalent base-eight numeral by repeated division. Study the following examples.

Example 1.

$$8\underline{)123}$$
$$8\underline{)15} \qquad R = 3 \qquad 3 \times 8^0$$
$$8\underline{)1} \qquad R = 7 \qquad 7 \times 8^1 \qquad 123_{(ten)} = 173_{(eight)}$$
$$0 \qquad R = 1 \qquad 1 \times 8^2$$

Example 2.

$$8\underline{)674}$$
$$8\underline{)84} \qquad R = 2 \qquad 2 \times 8^0$$
$$8\underline{)10} \qquad R = 4 \qquad 4 \times 8^1$$
$$8\underline{)1} \qquad R = 2 \qquad 2 \times 8^2 \qquad 674_{(ten)} = 1242_{(eight)}$$
$$0 \qquad R = 1 \qquad 1 \times 8^3$$

EXERCISE 2

1. Convert the following base-eight numerals to equivalent base-ten numerals.
   - (a) 144
   - (c) 1007
   - (b) 777
   - (d) 3276
2. Convert the following base-eight numerals to equivalent base-ten numerals.
   - (a) 1010
   - (c) 4066
   - (b) 6365
   - (d) 32,767
3. How many digits are used in the octal system?
4. What is the decimal value of "5" in each of the following octal numerals?
   - (a) 562
   - (c) 5637
   - (b) 53
   - (d) 51,677
5. Use the expanded notation method to change the following base-ten numerals to equivalent octal numerals.
   - (a) 777
   - (c) 8888
   - (e) 32,768
   - (b) 2323
   - (d) 9399
   - (f) 32,769
6. Use the repeated division method to convert the following decimal numerals to equivalent octal numerals.
   - (a) 1001
   - (c) 4745
   - (e) 4096
   - (b) 333
   - (d) 100
   - (f) 5238

7. Use either method discussed to convert the following decimal numerals to equivalent octal numerals.

(a) 345   (c) 4512   (e) 7100
(b) 1300   (d) 4056   (f) 72,000

6. BINARY OCTAL RELATION

Binary numerals have many digits and are clumsy. Since they contain so many digits it is hard to see at a glance what they represent.

Binary numerals convert very easily to octal numerals which contain fewer digits. Study the following examples.

Example 1.

$$1,011,110_{(two)} = \underline{(1 \times two^6)} + \underline{(0 \times two^5) + (1 \times two^4) + (1 \times two^3)}$$
$$+ \underline{(1 \times two^2) + (1 \times two^1) + (0 \times two^0)}$$

Each portion of the binary numeral separated by commas has been underlined in the expanded notation.

Notice that the first underlining marks the number represented by "1"; the second underlining, the number represented by "011"; and the third, the number represented by "110."

Observe the three underlined expressions.

$$\underline{(1 \times two^6)} = \text{sixty-four}$$
$$= 1 \times eight^2$$
$$\underline{(0 \times two^5) + (1 \times two^4) + (1 \times two^3)} = \text{zero} + \text{sixteen} + \text{eight}$$
$$= 0 + 16 + 8$$
$$= 24$$
$$= 3 \times eight^1$$
$$\underline{(1 \times two^2) + (1 \times two^1) + (0 \times two^0)} = \text{four} + \text{two} + 0$$
$$= 4 + 2 + 0$$
$$= 6$$
$$= 6 \times eight^0$$

Combining these, we have

$$1,011,110_{(two)} = 136_{(eight)}$$

Example 2.

$$11{,}111{,}001_{(two)} = (1 \times two^7) + (1 \times two^6)$$
$$+ (1 \times two^5) + (1 \times two^4)$$
$$+ (1 \times two^3) + (0 \times two^2)$$
$$+ (0 \times two^1) + (1 \times two^0)$$

$$\overline{(1 \times two^7) + (1 \times two^6)} = \text{one hundred twenty-eight}$$
$$+ \text{sixty-four}$$
$$= 128 + 64$$
$$= 192$$
$$= 3 \times 64$$
$$= 3 \times eight^2$$

$$\overline{(1 \times two^5) + (1 \times two^4) + (1 \times two^3)} = \text{thirty-two} + \text{sixteen} + \text{eight}$$
$$= 32 + 16 + 8$$
$$= 56$$
$$= 7 \times eight^1$$

$$\overline{(0 \times two^2) + (0 \times two^1) + (1 \times two^0)} = 0 + 0 + 1$$
$$= 1 \times eight^0$$
$$11{,}111{,}001_{(two)} = 371_{(eight)}$$

Using expanded notation makes the conversion from base two to base eight seem much more complicated than it really is. The binary numeration system is based on powers of two. Eight, the base of the octal system, is itself a power of two, $2^3 = 8$. Let's look again at the two examples just given.

$$1{,}011{,}110_{(two)} = 136_{(eight)}$$
$$11{,}111{,}001_{(two)} = 371_{(eight)}$$

Consider each of the groups of numerals, separated by the commas, as individual binary numerals and compare them with the following:

Base Two	Base Eight
1	1
10	2
11	3
100	4
101	5
110	6
111	7

Each group of three binary digits can be converted directly to its octal equivalent.

Observe the relation between the place values of the binary system and the octal system.

$$1_{(two)} = two^0 = eight^0 = 1_{(eight)}$$
$$1,000_{(two)} = two^3 = eight^1 = 10_{(eight)}$$
$$1,000,000_{(two)} = two^6 = eight^2 = 100_{(eight)}$$
$$1,000,000,000_{(two)} = two^9 = eight^3 = 1,000_{(eight)}, \text{ and so forth.}$$

From the foregoing we see that a regrouping, or a "carry," into each higher order of binary digits (separated by commas) forms the next higher power of eight.

Some other conversions are

$$111,101,011,100_{(two)} = 7534_{(eight)}$$
$$100,110,010,001_{(two)} = 4621_{(eight)}$$

Converting from octal to binary uses the same method.

Octal numeral	4	2	6	7
Binary numeral		100,010,110,111		

Therefore $4267_{(eight)} = 100,010,110,111_{(two)}$.

EXERCISE 3

1. Write the first twenty binary numerals starting with 1.
2. Write the first twenty octal numerals starting with 1.
3. Convert the following base-two numerals to base-eight numerals using expanded notation.
   - (a) 1,001
   - (b) 101,010
   - (c) 111,111
   - (d) 100,100,100
   - (e) 11,101,011
   - (f) 1,001,101
   - (g) 10,000,111
   - (h) 110,011,110
4. Use the binary-octal relation to convert each of the following binary numerals to base-eight numerals.
   - (a) 110,111
   - (b) 101,101,101
   - (c) 11,110,100
   - (d) 111,001,010
   - (e) 10,101,011
   - (f) 100,001,011
5. Use the binary-octal relation to convert each of the following octal numerals to binary numerals.
   - (a) 1234
   - (b) 7462
   - (c) 5731
   - (d) 36,245
   - (e) 77
   - (f) 100
   - (g) 56,200,063
   - (h) 44,766

6. Use the symbols "=," "<," ">" to compare the following pairs of numerals.

(a) $1,010_{(eight)}$, $268_{(eight)}$　　　　(d) $63_{(eight)}$, $1,011_{(ten)}$

(b) $100_{(ten)}$, $101_{(eight)}$　　　　(e) $111,000_{(two)}$, $100_{(ten)}$

(c) $11,111_{(two)}$, $25_{(eight)}$　　　　(f) $42_{(eight)}$, $101,010_{(two)}$

7. COMPUTERS AND THE OCTAL SYSTEM

As was stated previously, any character used by a digital computer, whether a number, a letter of the alphabet, or a symbol, must be represented by a combination of the two binary digits "0" and "1." For the purpose of representing a letter or symbol, the computer's words are subdivided into groups of binary digits (**bits**). The bits within each of these groups are treated as a unit and the group itself is called a **byte**.

The number of binary digits per byte determines the "working base" of the machine. The IBM 7094 has a 36-bit word and a six-bit byte. Each byte can represent one letter of the alphabet, one symbol, or *two* numbers. Therefore the 7094 uses six binary digits to represent one letter or symbol and three binary digits to represent a number. The working base of the 7094 is base eight. The program instructs the computer as to the type of data it is using (numbers or letters). Since the IBM 7094's word contains 36 binary digits, one word may represent twelve octal digits or six letters or symbols. Let us examine what one word might look like if it represented the integral numeral 222.

First convert 222 to base eight:

$$\begin{array}{r} 8\,\underline{|222} \\ 8\,\underline{|27} \quad R = 6 \\ 8\,\underline{|3} \quad R = 3 \\ 0 \quad R = 3 \end{array} \qquad 222_{(ten)} = 633_{(eight)}$$

Using the direct octal to binary conversion, we have

$$633_{(eight)} = 110,011,011_{(two)}$$

A representation of how this numeral might appear in a IBM 7094 word is shown in Figure 10.1.

figure 10.1

The IBM 7094 uses the following binary patterns for the first six letters of the alphabet.

Letter	Binary Pattern	Octal Equivalent
A	010,001	21
B	010,010	22
C	010,011	23
D	010,100	24
E	010,101	25
F	010,110	26

If these six letters were contained sequentially within one word, it would appear as shown in Figure 10.2.

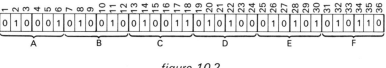

figure 10.2

8. COMPUTERS AND BASE SIXTEEN

The IBM System/360 uses an eight-bit byte. Therefore each letter or symbol is represented by eight binary digits and each number is represented by four binary digits. The 360's working base is **hexidecimal** or base sixteen.

The base sixteen numeration system has sixteen digits. The first ten digits are our familiar $0, 1, \ldots, 9$, and International Business Machines has chosen for the remaining six digits the first six letters of the alphabet. Thus:

$$A_{(sixteen)} = 10_{(ten)}$$
$$B_{(sixteen)} = 11_{(ten)}$$
$$C_{(sixteen)} = 12_{(ten)}$$
$$D_{(sixteen)} = 13_{(ten)}$$
$$E_{(sixteen)} = 14_{(ten)}$$
$$F_{(sixteen)} = 15_{(ten)}$$

A person using computers to their fullest capacity must be versed in many systems of numeration.

JOHN W. BACKUS (1925–   ) conceived the idea of FORTRAN (formula translation) computer language in 1954 while working on programming research at IBM. His effort to construct FORTRAN began with the recognition that the cost of programmers was as great as the cost of computers. Three years later, in 1957, the programming language FORTRAN was developed. Although only ten years old, FORTRAN is solidly established as the language of thousands of scientists and engineers who communicate with computers.

A Computer
Language/11

1. INTRODUCTION

Computers are wired to perform specific operations such as addition and multiplication. Since a computer does nothing on its own, there must be some way to tell it what to do and when to do it. A sequence of statements, called a **program** must be prepared to instruct the computer's operation. This sequence of instructions is written following certain rules determined by the language used. Just as the people of France speak French and the people of Germany speak German, each type of computer has its own unique language.

Early in computer development the shortcomings of a different language for each type of computer became evident. Although much time and effort was expended to develop a program, it could be used on only one type of computer. To remedy this, some universal languages were developed. It is possible for a program written in a universal language to operate on many computers.

One of these universal languages is FORTRAN IV. The word FORTRAN is a contraction of **formula translation**. This language is widely used throughout the computer world. In this chapter only the skeleton of the FORTRAN language will be discussed. The authors do not claim that reading this chapter will make a programmer of the reader. It is hoped that the material selected will help the reader understand one way in which a computer is directed to solve specific problems.

2. THE FORTRAN STATEMENT

One feature of FORTRAN is that the instructions contained within the program can be written to resemble closely the equations and formulas to be used. Each separate instruction is called a **statement**. After all statements necessary to solve a particular problem have been written in proper order, the statements are punched on cards. This deck of cards is submitted to the computer where another program, called a **compiler**, translates the FORTRAN statements into the language of the machine being used.

FORTRAN statements are written on **coding paper** (Figure 11.1). A typical card that statements are punched on is shown in Figure 11.2. Notice that the paper and card contain eighty columns. The statements themselves are written between columns seven and seventy-two inclusive. Columns one through five are used for statement numbers and column six is used for indicating the continuation of a statement started on the previous card.

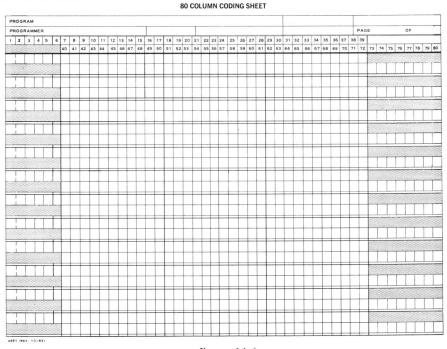

figure 11.1

3. CONSTANTS, VARIABLES, AND NAMES

A **constant** is a numerical value that appears in a FORTRAN statement. In the statements

$$I = 6 + J$$
$$A = 2.0 - B$$

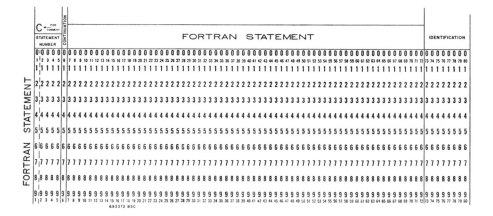

figure 11.2

the constants are 6 and 2.0. Two types of constants are used in FORTRAN, namely, **integer** and **real**. **Integer constants** are elements from the set I of integers, $I = \{\ldots, -2, -1, 0, 1, 2, \ldots\}$. In a FORTRAN statement an integer constant may *never* contain a decimal point. In the statement $I = 6 + J$, 6 is an integer constant. **Real constants** are elements from the set of real numbers, and *always* contain a decimal point when appearing in a FORTRAN statement. In the statement $A = 2.0 - B$, 2.0 is a real constant. Commas are *never* used in constants appearing in FORTRAN statements. For example, 20,000 is written 20000 and 13,456.8 is written 13456.8.

Examples of integer and real constants are given in Table 11.1.

TABLE 11.1

Integer Constants	Real Constants
1	11.2
-74	99.63942
22069	-0.9639
-11	-333.33
-7433	0.0
0	6269.7

In FORTRAN the term **variable** is used to denote any quantity that is referred to by name rather than as an explicit numerical value.

A variable may name many different values as the program executes. In the statement

$$\text{ALPHA} = 90.0 - \text{BETA}$$

ALPHA and BETA are variables.

In naming variables within a FORTRAN program, certain rules are followed. Any variable that represents an integer *must* be given a name that starts with one of the six letters: *I, J, K, L, M, N*. Real variable names may begin with any letter of the alphabet *except* those used for integer variables, that is, any letter except *I, J, K, L, M, N*. Names can contain one to six letters. With the exception of the starting character, names may contain numbers as well as letters. Some examples of variables assigned to constants are given in Table 11.2.

TABLE 11.2

I	= 2
ALPHA	= 10.6
SUM	= 2247.393
NAME	= −23
A33	= 100.0
KEY	= 44
ZERO	= 74.9
IGUESS	= 0

There is always a maximum and minimum value that may be defined by a constant or assumed by a variable. The numerical value of this maximum and minimum is determined by the construction of the machine but the range is always very large.

EXERCISE 1

1. Tell whether the following constants are integer or real.

(a) 30.6	(h) −392	(o) −333.3
(b) 6	(i) 63	(p) 0.23
(c) 0.0	(j) 32456	(q) 100
(d) 69.0	(k) 0.666	(r) −6302
(e) 13	(l) 0.723	(s) 19.111
(f) −409	(m) −6969	(t) −45.00
(g) 6834566	(n) 10.00	(u) 1567890

2. Select any appropriate variable name to assign to each of the following constants.

(a) −66.6   (d) −444   (g) 0.0

(b) 432   (e) 100.77   (h) −93

(c) 0   (f) 444269.2   (i) 536.9

3. Each of the following statements contains an error. Identify each error.

(a) INDEX = 22.3   (d) GAMMA = 16

(b) ACAT = 22,369.4   (e) IX = −1,673

(c) INTEGER = 2   (f) LAMBDA = −679.

4. Which of the following may name integer variables?

(a) JACK   (e) SQ66   (i) MNS101

(b) MEG   (f) PL401   (j) LPX

(c) SUZIE   (g) MUN86   (k) LZP

(d) KUK   (h) GO69   (l) I93

5. Which columns of FORTRAN coding paper may be used for statements?

6. Which columns on FORTRAN coding paper may be used for statement numbers?

7. Which column on a punched card is used for indicating the continuation of a statement stated on a previous card?

4. ARITHMETIC OPERATIONS

Five arithmetic operations are used in FORTRAN, namely: addition, subtraction, multiplication, division, and exponentiation (that is, raising a number to a power). These operations relate variables and constants within FORTRAN statements by the use of appropriate symbols. The symbols for these operations are given in Table 11.3.

TABLE 11.3

Symbol	Operation
+	addition
−	subtraction
*	multiplication
/	division
**	exponentiation

Some simple FORTRAN arithmetic statements and their meanings follow:

A = B + C

> The values assigned to variables B and C are added and the sum is assigned to variable A.

SUBTOT = SUM1 − SUM2

> The value of variable SUM2 is subtracted from the value of variable SUM1 and the difference is assigned to variable SUBTOT.

PROD = U1 * U2

> The values of variables U1 and U2 are multiplied and the product is assigned to variable PROD.

ANS = YEAR / 12.0

> The value of variable YEAR is divided by the constant 12.0 and the quotient is assigned to variable ANS.

AREA = SIDE ** 2

> The value of variable SIDE is squared and the answer assigned to variable AREA. We could have written this as AREA = SIDE * SIDE.

Since a single FORTRAN statement may contain a combination of arithmetic operations, we must know the order of operation. The rule is: exponentiation first, multiplication and division second, and addition and subtraction last. Let us examine some FORTRAN statements and their order of operation.

TOTAL = T1 + T2 + T3 + T4

> Variable TOTAL will be assigned the sum of the values of variables T1 + T2 + T3 + T4.

DEGREE = ANG + RAD * 57.29578

> The value represented by variable RAD will be multiplied by the constant 57.29578. To this product will be added the value of variable ANG. The results will be assigned to (stored in) variable DEGREE.

FIX = OLDANS − ANS / DEL ∗∗ 5   The value of variable DEL will be raised to the fifth power. The value of variable ANS will be divided by the fifth power of DEL and this quotient will be subtracted from the value of variable OLDANS. The result of the subtraction will be assigned to the variable FIX.

Sequences of the same operation are performed from left to right as they appear in the statement. Some examples follow.

Statement	Interpretation
A = A1 + A2 + A3	A = (A1 + A2) + A3
B = B1 − B2 − 6.0	B = (B1 − B2) − 6.0
C = C1 ∗ 80.0 ∗ C3	C = (C1 ∗ 80.0) ∗ C3
D = 77.0 / D1 / D2	D = (77.0 / D1) / D2

Now let us consider the equation

$$x = \frac{y}{a + b}$$

What FORTRAN statement or statements should we use to write this equation? One way to write this statement is

SUM = A + B   The sum of the value of A + B is saved in SUM.
X = Y / SUM   The value of Y is divided by the value of SUM and assigned to X.

The need for the statement SUM = A + B is apparent if we consider writing the equation as a single statement

$$X = Y / A + B$$

The order of operation is division first and addition second which translates, in algebraic notation, as

$$x = \frac{y}{a} + b$$

which is not equivalent to the given equation.

By using parentheses, operations of lower hierarchy can be grouped as though they were a single variable. This allows equations like the foregoing to be written in one FORTRAN statement:

$$X = Y / (A + B)$$

The parentheses say that variables A and B are added first and their sum divides Y. If more than one set of parentheses is used, the expression contained in the innermost set is computed first, then the second set, and so forth, working from the inside out.

In the FORTRAN language, the operation symbols $(+, -, *, /)$ cannot be used side by side. If we wish to multiply Y by the negative of X we cannot write $Y * -X$. Parentheses must be used to separate the operations thus: $Y * (-X)$.

Table 11.4 gives several FORTRAN statements and their equivalent algebraic equations. Notice the use of parentheses.

TABLE 11.4

FORTRAN Statement	Mathematical Equation
X = A − B + C	$x = a - b + c$
X = A − (B + C)	$x = a - (b + c)$
Z = A * B / C * D	$z = \dfrac{a \cdot b}{c} \cdot d$
Z = (A * B) / (C * D)	$z = \dfrac{a \cdot b}{c \cdot d}$
Y = A * B * C **2	$y = a \cdot b \cdot c^2$
Y = (A * B * C) **2	$y = (a \cdot b \cdot c)^2$
Z = X * A + B * X + C	$z = ax + bx + c$
Z = (X * A + B) * X + C	$z = ax^2 + bx + c$
X = A + B * A − (C − D) + B / E + C	$x = a + ba - (c - d) + \dfrac{b}{e} + c$
X = A + B * (A − (C − D) + B / (E + C))	$x = a + b\left[a - (c - d) + \dfrac{b}{e + c}\right]$
Z = B / (A + 32.0 / C + D)	$z = \dfrac{b}{(a + \dfrac{32}{c} + d)}$
Z = B / (A + 32.0 / (C + D))	$z = \dfrac{b}{a + \dfrac{32}{c + d}}$

Spaces are permitted in FORTRAN statements. The use of spaces is a matter of individual choice since they have no effect on the meaning of the statement. The statements

$$\text{FOVA} = 6.0 * (A + B) / 40.0$$
$$\text{FOVA} = 6.0 * (A + B)/40.0$$
$$\text{FOVA} = 6.0 * (A + B) / 40.0$$

are equivalent.

EXERCISE 2

1. Write an equivalent FORTRAN statement for each of the following.

    (a) $x = a + b(c - d^2)$ (b) $x = a + \dfrac{a \cdot c}{b \cdot d}$

2. Write an equivalent FORTRAN statement for each of the following.

    (a) $y = \dfrac{1.0 - \dfrac{a \cdot b}{c \cdot d}}{1.0 + \dfrac{a \cdot b}{c \cdot d}}$ (b) $y = \left[\dfrac{a - (b + c)}{16.0} \right] a \cdot b^2$

3. Write an equivalent FORTRAN statement for each of the following.

    (a) $y = \dfrac{a \cdot b}{2.0} + \dfrac{c \cdot d}{2.0}$ (b) $y = \dfrac{a \cdot b}{1.0 + \dfrac{c}{d}}$

4. Write an equivalent FORTRAN statement for each of the following.

    (a) $z = \left[\dfrac{(a \cdot b)^x}{(c \cdot d)^y} \right]^2$ (b) $z = \dfrac{a^2}{1.0 + \dfrac{\dfrac{a}{b}}{\dfrac{c}{d}}}$

5. Write an equivalent mathematical equation for each of the following FORTRAN statements. Letters may be substituted for variable names.
    (a) AREA = 3.1415 * R **2
    (b) ANS = (A / GEE + B / GEE)
    (c) X2 = ((A * B) / (C * D)) **2

6. Write an equivalent mathematical equation for each of the following FORTRAN statements. Letters may be substituted for variable names.
    (a) ADJUST = A − (A / A **X)
    (b) ANINC = 1.0 / (17.0 + (A / B))
    (c) RES = ALPHA * 2.0 / EXP * VAL − ALPHA * 2.0 / (EXP * VAL)
    (d) TOT = ((A / B * C) ** 2 / ((B * C) / A) ** 2) ** 4

7. In the following examples, let A = 2.0, B = 3.0, C = 1.0, and D = 4.0. Evaluate the FORTRAN statements for X.
   (a) X = 2.0 * A − (B + C)   (b) X = (A + C) / (B − D)
8. In the following examples, let A = 4.0, B = 6.0, C = 5.0, and D = 2.0. Evaluate the FORTRAN statements for X.
   (a) X = B + D − A **2   (b) X = A + C / D − B
9. In the following examples, let A = 10.0, B = 5.0, C = 15.0, and D = 5.0. Evaluate the FORTRAN statements for X.
   (a) X = D / (A / (B + C))   (b) X = ((A + D) / B) **3
10. In the following examples, let A = − 2.0, B = − 7.0, C = − 2.0, and D = 3.0. Evaluate the FORTRAN statements for X.
    (a) X = ((A + D) / (A − D)) / (−A)
    (b) X = (D + 5.0 * A) / ((C **2 + 2.0) / 2.0 * B)

5. INTEGER ARITHMETIC

The arithmetic operations of addition, subtraction, and multiplication performed on integers pose no problem in FORTRAN statements. For example,

if I = 6 + 9	then I = 15
if I = 4 + (− 7)	then I = − 3
if I = 32 − 2	then I = 30
if I = − 6 − (− 20)	then I = 14
if I = 7 * 5	then I = 35
if I = − 3 * 6	then I = − 18

Division of integers, on the other hand, may create a special problem. The division of one integer by another does not always result in an integral quotient. *When two integers are divided in FORTRAN, the fractional part (the remainder) is truncated (that is, discarded).* For instance, if we divide 6 by 4 the result is 1.5. In FORTRAN, 6/4 would result in the quotient 1 with the remainder 0.5 discarded. Study the following examples of FORTRAN division.

I = 5 / 3 * 6	I = 6
I = 6 * 5 / 3	I = 10
I = 3 / 4 * 2	I = 0
I = 8 / 4 * 12	I = 24
I = 12 / 8 * 4	I = 4

EXERCISE 3

Determine the value of I in each of the following.

1. $I = 16 / 4 * 2$
2. $I = 4 * 3 / 7$
3. $I = 7 + 9 / 4$
4. $I = (9 + 6) / 5$
5. $I = 2 * 6 / 3 * 8$
6. $I = 2 + 3 / 4 + 5$
7. $I = (2 * 6) / (3 * 8)$
8. $I = (2 + 3) / (4 + 5)$

9. $I = 8 / 3 * 5 / 9$
10. $I = 17 / (8 * 2)$
11. $I = 6 * 7 / (5 * 9)$
12. $I = 45 / 8 * (17 / (2 * 8))$
13. $I = (16 + 4 * 6 * 3) / 20$
14. $I = ((3 + 5) * 7 / 5) / 2 * 3$
15. $I = (6 * 9) / 2 / 2$
16. $I = ((13 - 3) * 6 - 3 / 2) / 2$

6. TRANSFER STATEMENTS

In the previous sections of this chapter we studied the composition of the arithmetic statements of the FORTRAN program. We now discuss **transfer statements**. Much of the power of programming is the result of transfer statements. Transfer statements allow certain parts of the program to be used and other parts to be skipped. The transfer statements we shall study can be classified into three groups: **unconditional transfers, arithmetic transfers**, and **logical transfers**.

A transfer statement refers to specific statement numbers. Statement numbers must be written and punched in columns 1 through 6 of the FORTRAN coding forms and cards. These numbers are names of the statements they represent. Statement numbers must not be duplicated within a program.

The **unconditional** transfer:

GO TO n

means exactly what it says: Go next to execute the statement whose name is n. For example:

GO TO 6   Go next to the statement whose name is 6
GO TO 3   Go next to the statement whose name is 3

The arithmetic and logical transfers we shall discuss are in the form of IF statements. **IF statements** are **conditional transfers**, that is, they may or may not skip parts of the program, depending on the condition or value of specified variables. With logical transfers, decisions are based on a logical quantity being true or false in value; with arithmetic transfers decisions are based on an arithmetic quantity being either less than zero, zero, or greater than zero.

The form of the **arithmetic transfer** is

$$\text{IF (a) } n_1, n_2, n_3$$

The arithmetic expression *a* may be a single variable or a series of arithmetic operations that will compute to a single result within parentheses. The expressions n_1, n_2, and n_3 are statement numbers. They must be separated by commas in the FORTRAN statement.

The arithmetic IF statement

$$\text{IF (a) } n_1, n_2, n_3$$

means: if the value of the arithmetic expression *a* is less than 0, execute next the statement named n_1; if the arithmetic expression *a* is 0, then execute next the statement named n_2; if the arithmetic expression *a* is greater than 0, then execute next the statement named n_3.

Some examples of arithmetic IF statements and their meanings follow.

Arithmetic IF Statements	Meanings
IF (Z) 10, 20, 30	If Z is less than 0, go to statement 10; if Z is 0, go to statement 20; if Z is greater than 0, go to statement 30.
IF (BAL − WTHDRW) 1000, 500, 20	If BAL − WTHDRW is less than 0, go to statement 1000; if BAL − WTHDRW is 0, go to statement 500; if BAL − WTHDRW is greater than 0, go to statement 20.
IF (INC − IX / 5) 100, 200, 300	If INC − IX / 5 is less than 0, go to statement 100; if INC − IX / 5 is 0, go to statement 200; if INC − IX / 5 is greater than 0, go to statement 300.

A **logical transfer** is in the form

$$\text{IF (ex) ST}$$

where *ex* is a logical expression. A **logical expression** is an assertion (for example, *a* is greater than *b*). It is characterized by the fact that at any given time it can have two possible values: true or false. "ST" will be either an unconditional "GO TO" or an arithmetic statement.

The logical transfer

$$IF (ex) ST$$

means: if the logical expression *ex* is true then execute the statement ST and proceed; if the logical expression *ex* is false do not execute ST, simply proceed to the next statement.

The expression *ex* can relate two values by using one of six relational operators. The six **relational operators** used in logical transfer statements are given in Table 11.5.

TABLE 11.5

Relational Operator	Mathematical Symbol	Definition
.GT.	$>$	Greater than
.GE.	$\geq$	Greater than or equal to
.EQ.	$=$	Equal to
.NE.	$\neq$	Not equal to
.LE.	$\leq$	Less than or equal to
.LT.	$<$	Less than

Periods on either side of the relational operators are used to separate them from the variable names or constants. Study the following examples. Notice that care must be taken not to mix real and integer expressions within parentheses.

Logical IF Statements	Meanings
IF (A .GT. 12.0) B = 90.0 − A	If A is greater than 12.0, then execute B = 90.0 − A; If A is less than or equal to 12.0 then go to the next statement.
IF (I .EQ. 20) GO TO 300	If I is equal to 20, then execute the statement named 300; if I is not equal to 20, then go to the next statement.
IF (I .NE. 1) M = N **2	If I is not equal to 1, then execute the statement M = N **2; if I is

equal to 1, then go to the next statement.

IF (GEE .GT. 32.2) GO TO 50

MACH = 0

GO TO 100

50 MACH = 1

If GEE is greater than 32.2, then go to statement 50 where MACH is set equal to 1; if GEE is less than or equal to 32.2, then execute the next statement where MACH is set equal to 0 and then go to statement 100.

EXERCISE 4

In the following examples, determine what will happen under the conditions given (problems 1–10).

1.   IF (2 − 3) 100, 200, 300

2.   IF (7.0 * X) 2, 4, 6   X = 0.0

3.   IF (N .GT. M) X = 70.0   N = 2, M = 0

4.   IF (ZEE * 3.0 .GE. 35.0) GO TO 20   ZEE = 15.0

5.   IF (ICOUNT .LE. 20) GO TO 40   ICOUNT = 21
        ICOUNT = ICOUNT − 1

6.   IF (IND) 10, 20, 30   IND = 2
     10 X = Y   Y = 10.0
        GO TO 40
     20 X = 2.0 * Y
        GO TO 40
     30 X = Y **2

7.   IF (W .LE. 0.0) GO TO 20   W = − 32.0
        IPOS = 1
        GO TO 30
     20 INEG = − 1

8.   IF (X * Y **2 − 72.0) 15, 40, 25   X = 2.0
     15 X = X + 1.0   Y = 6.0
        GO TO 40
     20 X = X − 1.0
        GO TO 45

9.   IF (X − Z * (16.0 − Y) .LT. 25.0) X = X **2
                                     X = 15.2
                                     Y = 5.5
                                     Z = 1.5

10. IF (7.GT. IX) GO TO 40   IX = 7
     CALC = FER **3   FER = 5.0
     GO TO 30   Z = 2
40   CALC = FER **Z
30   ...

Write statements without logical IF's (use arithmetic IF's) to duplicate the following (problems 11–15).

11.   IF (SUDS .LE. BIG) SOAP = X
12.   IF (NIGHT .EQ. 28) GO TO 10
         MOON = IOFF
         GO TO 20
10 MOON = ION
20   ...
13.   X = 65.0
         IF (IF .GE. − 5) GO TO 5
    3 X = X − 2.0
         GO TO 7
    5 X = X + 2.0
    7 SUM = SUM + X
14.   IF (ME .LT. 0) M = 0
         M = M + 1
15.   IF (HRS .GT. 40.0) GO TO 75
         PAY = RATE * 40.0
         GO TO 100
   75 PAY = RATE * 40.0 + (HRS − 40.0) * RATE * 1.5
  100   ...

7. WRITING PROGRAMS

The addition of four more statements, READ, WRITE, STOP, and END, gives us enough tools to write small programs. The **READ** and **WRITE** statements instruct the computer to "read" in data and "write" out the desired answers. The complete explanations of READ and WRITE statements are beyond the scope of this book. They are used here only to give the reader the idea that data must be brought in for processing, and answers must be written out. The form we shall use is

READ I, X, Y, Z

meaning: the actual numerical values of I, X, Y, and Z are to be assigned to the variables I, X, Y, and Z used in the program; and

WRITE SUBTOT, TOTAL, YEAR

meaning: print the numerical values of SUBTOT, TOTAL, and YEAR as computed by the program. In both READ and WRITE statements the variables must be separated by commas.

The statement **STOP** signals the computer to terminate the execution of the program.

The statement **END** must be the last statement of all FORTRAN programs. Since a program is initially punched on a deck of cards, with each statement placed on a separate card or group of cards, the card containing the END instruction must be physically the last card in the program deck.

When beginning a new program it is often helpful to diagram the actual steps to be taken. Such a diagram is called a flow chart. A **flow chart** is actually a list of steps to be performed in the sequence. It is customary to use the shapes in Figure 11.3 in a flow chart. Each shape has a particular function, as shown in Figure 11.3.

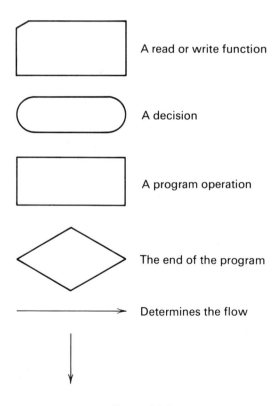

figure 11.3

Suppose we wish to write a program that would square each number in a set of numbers, sum the squares, and print out the sum. A flow chart for this program is shown in Figure 11.4. The actual

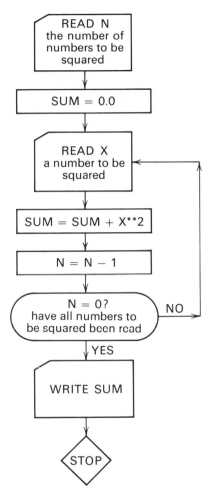

figure 11.4

FORTRAN statements of this program are written on FORTRAN **coding paper**, as shown in Figure 11.5. The statements in the program mean the following:

The actual numerical value of N is read (called **input**) and assigned to the variable N (the number of numbers to be squared).
Zero is assigned to variable SUM.

```
       C FOR COMMENT
STATEMENT  Cont.            FORTRAN STATEMENT
 NUMBER
1        5 6 7    10      15       20       25       30       35
             READ N
             SUM  =  0.0
    10       READ X
             SUM  =  SUM  +  X**2
             N  =  N-1
             IF(N.NE.0)GOTO10
             WRITE SUM
             STOP
             END
```

figure 11.5

The actual numerical value of X is read (called **input**) and assigned to
the variable X.

The value of the variable X is squared and added to SUM. The result
is assigned to the variable SUM.

One is subtracted from N and assigned to the variable N.

If N is not equal to 0, GO TO statement 10.

Print the numerical value of the variable SUM (this is called **output**).

STOP.

END.

In this case, each new X is read into the same core location as the previous X.
Notice that SUM was set equal to zero before the addition began. The
reason for this is that we have no idea what may have been left in the core
location assigned to SUM by another program.

	NUM	B	H	AREA
READ N	1	7.0	7.0	24.50
NUM=0	2	4.0	6.0	12.00
10 NUM=NUM+1	3	2.0	5.0	5.00
READ B,H	4	6.4	2.3	7.36
AREA=B*H/2.0	5	3.5	6.4	11.20
WRITE NUM,B,H,AREA				
IF(NUM.LT.N)GOTO10				
STOP				
END				

figure 11.6

Statements such as $SUM = SUM + X **2$ and $N = N - 1$ cause the value of the word referenced to be modified by the operation indicated. The modified value of the variable replaces the original value in core storage.

The formula for the area of a triangle is $b \cdot h/2$ where b is the measure of the base and h is the measure of the height. Some arbitrary values for b and h follow:

n	b	h
1	7.0	7.0
2	4.0	6.0
3	2.0	5.0
4	6.4	2.3
5	3.5	6.4

Figure 11.6 shows a program to find the area of a triangle. Included are the answers obtained using the foregoing values. The statements in this program mean:

The actual numerical value of N is read as input and assigned to variable N (the number of areas to be computed).

Variable NUM is given a zero value.

The value of the variable NUM is increased by 1; this is used as a counter.

The actual numerical values of B and H are read as input and assigned to variables B and H respectively (the base and height of a triangle).

B is multiplied by H, the product is divided by 2.0, and the result is saved in variable AREA.

Print as output the numerical values of the variables; NUM (the counter), B (the base of the triangle), H (the height of the triangle), and AREA (the area of the triangle).

Is the value of NUM less than N? If it is not we have not completed the required number of iterations and we proceed to statement number 10.

If NUM is equal to or greater than N then the process is completed and the program executes STOP.

The arithmetic average, or mean,* of a set of numbers is computed by adding all the numbers of the set and dividing by the number of numbers

* See Chapter 7, Section 6.

in the set. The following is a FORTRAN program to find the mean of a set of *n* numbers.

READ N	Read the number of numbers to be averaged.
SUM = 0.0	Set the cell used for totaling equal to zero.
I = 0	Set a counter equal to zero.
1 READ SAMPLE	SAMPLE is a member of the set.
SUM = SUM + SAMPLE	Add the new member of the set to the others.
I = I + 1	Increment the count of the number of samples read.
IF (I .NE. N) GO TO 1	If we have not read *n* samples go back for more.
RN = N	Convert integer N to a real variable.*
AVE = SUM / RN	Compute average.
WRITE AVE	Print the answer.
STOP	
END.	

EXERCISE 5

1. Draw a flow chart for the sample problem in Section 7 which found the areas of triangles.
2. Draw a flow chart for the sample problem to find the average of a set of *n* numbers.
3. Draw a flow chart showing the steps needed to find the greatest number in a set of *n* numbers.
4. Write a program for the exercise in problem 3.
5. Draw a flow chart showing the steps needed to compute and compare the mean of a set of *n* numbers with the mean of the greatest and least values in the set.
6. Write a program for the exercise in problem 5.

* In Section 3 of this chapter we indicated equating integer and real values could be in error. FORTRAN will automatically convert the expression on the right of the equal sign to the type of variable indicated on the left of the equal sign.

7. What is being computed in the following?

$$PI = 3.146$$
$$READ\ R$$
$$AREA = PI * R ** 2$$
$$WRITE\ AREA$$
$$STOP$$
$$END$$

8. What would be the values of A and I in the following cases if N = 4 and B = 23.6?
   (a) A = N   (b) I = B

9. Draw a flow chart and then write the program to find the mean and standard deviation* of a set of five numbers. For the square roots needed, use an exponent of 0.5† that is SQRA = A **.5.

8. DIMENSION STATEMENTS AND SUBSCRIPTS

Each constant and variable that we have studied is assigned to a single word of core storage. As the compiling program scans the FORTRAN statements, each unique name is assigned a particular word in core. In the statement

$$SUM = X + 12.0$$

the values which the names SUM, X, and 12.0 represent are each in different words in core.

More than one core location can be assigned to a variable name by using a **DIMENSION statement.** These statements are written in the form

DIMENSION declaration

The declaration tells the number of variables to which more than one core location is to be allocated. The variable name is followed by the number of locations to be assigned. The number is always enclosed in parentheses. For example,

DIMENSION A(20)

Variable A is assigned twenty words. The twenty words are contiguous in core and are called an **array**. Since variable A is an array of twenty words, we must have a means of referring to each particular element of the array.

* See Chapter 7, Section 10, for the formula for the standard deviation.

† $\sqrt{a} = a^{\frac{1}{2}} = a^{0.5}$.

If we wish to refer to the second element of array A, we would write A(2). The (2) is called the **subscript**. The tenth word of the array is A(10), the first word A(1), and so forth. Subscripts denote a particular element of an array and are always enclosed in parentheses.

If a dimension is assigned to more than one variable, their names must be separated by commas in the DIMENSION statement. For example,

$$\text{DIMENSION P(3), Q(4)}$$

means variable P has three words, P(1), P(2), P(3), assigned to it, and variable Q has four words, Q(1), Q(2), Q(3), Q(4), assigned to it. Figure 11.7 is a pictorial representation of the arrays P and Q in core.

P(1)	P(2)	P(3)	Q(1)	Q(2)	Q(3)	Q(4)

figure 11.7

The value represented by Q(1) occupies the core location immediately following the value represented by P(3). Care must be taken that subscripts do not refer to elements outside their own array.

Each of the foregoing variables have one subscript and are called **singularly subscripted variables**. In FORTRAN, variables may have more than one subscript. The maximum number of subscripts that a variable may have depends on the computer used, but it is never less than three. We shall limit our variables to two subscripts, called **doubly subscripted variables**.

The number of core locations to be allotted to doubly subscripted variables is also defined in a DIMENSION statement. For example, the statement

$$\text{DIMENSION X(2, 2), A(4)}$$

means that X(2, 2) is a $2 \times 2 = 4$ element array. The order in which they appear in core is shown in Figure 11.8.

X(1,1)	X(2,1)	X(1,2)	X(2, 2)	A(1)	A(2)	A(3)	A(4)

figure 11.8

The DIMENSION statement is the *first* statement of a FORTRAN program. This is to establish the sizes and types of arrays before they are used in arithmetic or transfer statements.

EXERCISE 6

1. Write the DIMENSION statements for the following.
   (a) Array A has six elements.
       Array B has four elements.
       Array C has ten elements.
   (b) Array X has seventeen elements.
       Array I has twenty elements.
   (c) Array P has ten elements.
       Array R is a 2 × 2 array.
       Array S is a 3 × 3 array.
2. Draw a figure representing the core layout of a three-element array Y, followed by a 2 × 2 array Z.
3. How many core locations would be needed for the doubly subscripted array TEM (10, 20)?
4. Draw a figure representing the arrangement of the elements of M(3, 2) in core.
5. Using DIMENSION AF(30), ZEE(10, 5), M(4, 6).
   (a) What is the total number of cells used by the three arrays?
   (b) What variable name immediately follows AF(30)?
   (c) What variable name immediately follows ZEE(6, 2)?
   (d) What variable name immediately precedes ZEE(1, 4)?
   (e) What variable name is contained in the tenth cell after ZEE(5, 5)?
6. Write the names of the first, fifth, and eleventh elements of array Z(60).
7. Write the names of the third, twenty-fifth, and thirty-second elements of array HIPPY(20, 3).
8. Name the subscripts of the following variables.
   (a) SUZIE(7)   (b) GAI(2, 5)   (c) MEG(1)   (d) ATLAS(13)
9. How many core locations are needed for the statement.
       DIMENSION JOHN(7, 11), DOGG(102), ETT(60)

9. THE DO STATEMENT

Probably the most important statement in FORTRAN is the **DO statement**. Each DO statement causes an iteration to be performed. The form of the DO statement is

$$\text{DO ST IX} = I_1, I_2$$

where ST is a statement number. The DO statement causes all statements between it and statement ST, including ST, to be executed repeatedly.

The number of times each statement is repeated is controlled by IX, I_1, and I_2. IX is a counter or **index**. Starting at its initial value of I_1, IX is increased by one each iteration until it reaches the value I_2, where the repetition is terminated. IX must be a name of a nonsubscripted integer variable. I_1 and I_2 can be either names for nonsubscripted integer variables or integer constants. Neither I_1 nor I_2 can assume a zero or negative value. Notice a comma separates the limiting names.

One DO statement is

$$DO\ 25\ I = 1, 3$$

This DO statement causes all the subsequent statements up to and including statement 25 to be repeated three times. In the three repetitions I is assigned the values 1, 2, and 3 respectively.

In the statement

$$DO\ 60\ N = 1, M$$

all statements up to and including 60 will be repeated M times, M having been equated previously to a positive integer greater than zero. In each iteration N is assigned the number indicating the iteration.

In the statement

$$DO\ 100\ I = J, K$$

all statements up to and including 100 will be executed $K - J + 1$ times. Both J and K have been assigned positive integer values greater than zero in the previous statement. In the first iteration I is assigned the value of J. I is incremented by 1 on each subsequent iteration until the iteration in which $I = K$ is completed. Then the repetition is terminated and the program continues by executing the statement immediately after 100. In this case $J \le K$.

We shall always use the statement CONTINUE as the last statement in the DO range. The continue statement will have the statement number ST. An example is:

$$DO\ 5\ I = 1, 3$$

$$\vdots$$

$$5\ CONTINUE$$

All statements between and including the DO and CONTINUE statements comprise what is called a **DO loop**. The word *loop* is used because the program loops through the indicated statements until the value of the index has reached the maximum value.

Statements within a DO loop may contain array names subscripted by the index of the DO statement. This allows the program to operate on all or part of the elements of arrays; the array names must have been mentioned previously in a DIMENSION statement.

Suppose we want to set all elements of an array named MEG (which has fifteen words) equal to zero. This is done in the following manner:

```
DIMENSION MEG(15)
DO 10 I = 1, 15
MEG(I) = 0
10 CONTINUE
```

The dimension statement causes fifteen core locations to be assigned to the name MEG. In the first iteration, $I = 1$, and $MEG(1) = 0$; in the second iteration, $I = 2$, and $MEG(2) = 0$; and so forth. This process continues until the last iteration, where $I = 15$, is reached and $MEG(15) = 0$.

Suppose we wish to store the first one hundred positive integers in an array INTGER(200). At the same time we wish to store in array INTSUM(200) the first one hundred sums of the positive integers—that is, cell INTSUM(5) equals the sum of $1 + 2 + 3 + 4 + 5$, cell INTSUM(N) equals the sum of $1 + 2 + 3 + \cdots + N$, and so on.

This may be written:

```
DIMENSION INTGER(200), INTSUM(200)
ISUM = 0
DO 20 N = 1, 100
INTGER(N) = N
ISUM = ISUM + N
INTSUM(N) = ISUM
20 CONTINUE
```

In the foregoing example the value of the index is used both as a subscript and as an integer variable.

In no case may a statement be given that will transfer program control to a statement within a DO loop. The following example gives a case of illegal transfer.

```
DIMENSION A(30), I(2)
IF (I(1)) 5, 5, 10
5 DO 20 J = 1, 25
FJ = J
10 A(J) = FJ
20 CONTINUE
```

In this example the transfer to statement 10 is executed before the value of index J is established, thus causing an error.

An example of a program that finds the arithmetic mean and standard deviation* of n numbers stored in an array named S is given below. Array S has been given the dimension 200 which imposes the condition $n \leq 200$. A cell named N contains the value of n.

```
DIMENSION S(200)
SUM = 0.0
DO 20 L = 1, N
SUM = SUM + S(L)
20 CONTINUE
FN = N
AM = SUM / FN
SUMSQ = 0.0
DO 40 L = 1, N
SUMSQ = SUMSQ + (S(L) − AM)**2
40 CONTINUE
SD = (SUMSQ/FN) **.5†
```

The variables AM and SD contain the arithmetic mean and standard deviation respectively.

EXERCISE 7

1. Give the meaning of each FORTRAN statement in the program for finding the arithmetic mean and standard deviation of a set of n numbers.

In the following problems (2–8), choose a compatible dimension size and variable names if these are not given.

2. Write the statements needed to sum the first sixteen numbers in an array named FP. Assign a total of fifty cells to array FP. Assign the sum to variable SUM.

3. Write the statements needed to sum the fifth to the ninth numbers in array NUM.

4. Write the statements necessary to find the largest number of a set of forty numbers.

* See Chapter 7, Section 10, for the formula for the standard deviation.

† $\sqrt{a} = a^{\frac{1}{2}} = a^{0.5}$.

5. Write the statements needed to find the arithmetic mean of a set of numbers.

6. Write the statements necessary to find how many numbers in array FIX(30) have a numerical value less than 100.0.

7. Write the statements necessary to find the maximum and minimum values in an array of *n* numbers.

8. Write the statements needed to find the arithmetic means of both the negative and positive integers contains in an array N(300). No value is equal to zero.

9. Determine what problem is being solved in the following.

    (a)   DIMENSION FAT(250)

```
            FMIN = 0.0
            DO 20 I = 1, 250
            IF (FAT(I).GE. FMIN) GO TO 20
            FMIN = FAT(I)
         20 CONTINUE
```

    (b)   DIMENSION SAMP(55)

```
            SUM = 0.0
            SUMSQ = 0.0
            DO 40 K = 1, 55
            SUM = SUM + SAMP(K)
            SUMSQ = SUMSQ + SAMP(K) **2
         40 CONTINUE
```

Answers to
Selected Problems

CHAPTER 1   EXERCISE 1

1. (a) T; (c) F; (d) F; (e) T; (f) T; (h) F; (i) F.
2. (a) Six is not divisible by 3.
   (b) John does not get A grades.
   (c) All rational numbers can be graphed on the real number line.
3. (a) Some talkative girls are not popular.
   (b) $\sqrt{2}$ is a real number.
   (c) All career women are beautiful.
4. (a) Doris drinks wine.
   (b) No parfaits are peeps.
   (c) All modular systems are groups.
5. (a) Meg has two ears.
   (b) No freshmen are handsome men.
   (c) Some college professors are not excellent talkers.
6. Valid.
7. Invalid.
9. (a) Invalid; (b) Invalid.

CHAPTER 1   EXERCISE 2

2. (a) $p \wedge q$; (b) $p \vee q$; (c) $\sim p$.
4. (a) T; (b) F; (c) T; (d) F.
7. (a) $p \rightarrow q$; (b) $q \rightarrow p$; (c) $p \leftrightarrow q$; (d) $q \rightarrow (\sim p)$; (e) $(\sim q) \rightarrow (\sim p)$.
8. (b); (c); (d); (e).
9. (a); (c); (d); (e).
10. (a) p; (b) q; (c) $p \wedge q$; (d) $q \wedge (\sim p)$; (e) $q \rightarrow p$; (f) $(\sim p) \wedge (\sim q)$.
12. (a) p; (b) q; (c) $p \rightarrow q$; (d) $(\sim q) \rightarrow p$; (e) $q \vee p$.

261

CHAPTER 1   EXERCISE 3

2. | p | q | r | $(q \lor r)$ | $p \to (q \lor r)$ |
    |---|---|---|---|---|
    | T | T | T | T | T |
    | T | T | F | T | T |
    | T | F | T | T | T |
    | T | F | F | F | F |
    | F | T | T | T | T |
    | F | T | F | T | T |
    | F | F | T | T | T |
    | F | F | F | F | T |

9. | p | q | $(p \land q)$ | $(p \land q) \to p$ |
    |---|---|---|---|
    | T | T | T | T |
    | T | F | F | T |
    | F | T | F | T |
    | F | F | F | T |

12. (a) (1) If n is divisible by 1, then n is an integer.
        (2) If n is not an integer, then n is not divisible by 1.
        (3) If n is not divisible by 1, then n is not an integer.
    (c) (1) If prices go down, then there is a depression.
        (2) If there is not a depression, then prices do not go down.
        (3) If prices do not go down, then there is not a depression.

13. | p | q | $(\sim p)$ | $(\sim q)$ | $(p \land q)$ | $(\sim p) \land (\sim q)$ | $\sim (p \land q)$ |
    |---|---|---|---|---|---|---|
    | T | T | F | F | T | F | F |
    | T | F | F | T | F | T | T |
    | F | T | T | F | F | T | T |
    | F | F | T | T | F | T | T |

18. Yes.
19. True.
20. (a) T; (b) F; (c) F; (d) T.

CHAPTER 1   EXERCISE 4

1. (a) Valid; (b) Valid.
2. (a) Invalid; (b) Valid.
3. (a) Invalid; (b) Invalid.
4. (a) Valid; (b) Valid.
5. (a) Invalid; (b) Valid.
6. (a) It is 8 o'clock P.S.T. in San Francisco.
   (b) 1972 is an election year.
7. (a) X was at the scene of the crime.
   (b) $ABCD$ is a rectangle.
8. (a) x is less than y.
   (b) Frank has two heads.
9. (a) n is an integer.
   (b) $x > z$.

CHAPTER 1   EXERCISE 5

1. $(p \lor q) \to (r \land q)$ is true. p and q are true, therefore $(p \lor q)$ is true by the definition of disjunction. By the law of detachment $(r \land q)$ is true. Since $(r \land q)$ is true, r is true by the definition of a conjunction.

3. $(p \lor q) \to r$ is true. Since both p and q are true, $(p \lor q)$ is true. Therefore, by the law of detachment, r is true.

5. p: This is a good course.
   q: This course is worth taking.
   r: Math is easy.

   $p \to q$
   $r \lor (\sim q)$
   $\sim r$
   _____
   $\therefore \sim p$

   Let us assume that $(\sim p)$ is false. Then p is true. Since $p \to q$ is true and p is true, q is true by the law of detachment. Since q is true, $(\sim q)$ is false. Since $(\sim q)$ is false and $r \lor (\sim q)$ is true, r is true. We now have a contradiction: r and $(\sim r)$ true. Hence our assumption is false and $(\sim p)$ is true.

7. p: Carl is elected class president.
   q: Bill will be elected vice-president.
   r: Betty will be elected secretary.

   $p \to q$
   $q \to (\sim r)$
   _____
   $\therefore r \to (\sim p)$

   Since $p \to q$ and $q \to (\sim r)$ are true, $p \to (\sim r)$ is true by the law of syllogisms. Since $p \to (\sim r)$ is true its contrapositive $r \to (\sim p)$ is true.

9. p: Vincent water skiis.
   q: It is summer.
   r: Blanche is in town.

   $p \to q$
   $r \to p$
   $\sim q$
   _____
   $\therefore \sim r$

   The contrapositives of the given implications are true. Hence $(\sim q) \to (\sim p)$, and $(\sim p) \to (\sim r)$ are true. By the law of syllogisms $(\sim q) \to (\sim r)$ is true. Since $(\sim q) \to (\sim r)$ is true and $(\sim q)$ is true, $(\sim r)$ is true by the law of detachment.

11. p: John is a thief.
    q: Newton is a shoplifter.
    r: Carl is guilty of car theft.

    p
    q
    $(\sim p) \to r$
    $q \to (\sim r)$
    _____
    $\therefore (\sim r)$

    Since $q \to (\sim r)$ and q are true, $(\sim r)$ is true by the law of detachment.

13. p: This class is a bore.
    q: The instructor is interesting.
    r: The subject is worthwhile.
    $p \rightarrow (\sim q)$
    $(\sim p) \rightarrow r$
    q

    $\overline{}$

    $\therefore r$

The contrapositive, $q \rightarrow (\sim p)$, of $p \rightarrow (\sim q)$ is true. By the law of syllogisms we now have $q \rightarrow r$. Since q is true and $q \rightarrow r$ is true, r is true by the law of detachment.

CHAPTER 2   EXERCISE 1

1. (a) 48×1; (b) 24×2; (c) 16×3; (d) 12×4; (e) 6×8; (f) 8×6.
2. (a) $1, 2, 3, 4, 6, 12$; (b) $1, 2, 4, 7, 8, 14, 28, 56$; (c) $1, 2, 3, 4, 6, 8, 12, 16, 24, 32, 48, 96$; (d) $1, 2, 3, 4, 6, 8, 9, 12, 16, 18, 24, 36, 48, 72, 144$.
3. $+ 1$.
4. 0.
5. (a) $1, 2$; (b) $1, 3$; (c) $1, 11$; (d) $1, 23$; (e) $1, 41$; (f) $1, 53$; (g) $1, 83$; (h) $1, 101$.
6. (a), (b), (c), (e), (g), (h).
7. $1, 2, 3, 4, 6, 8, 12$.
10. 2.
11. 1.
12. 0.
13. (a) $17 = (2 \times 8) + 1$; (b) $- 86 = 2(-42)$.

CHAPTER 2   EXERCISE 2

1. (a) $26 = (8)(3) + 2$; (b) $39 = (7)(5) + 4$; (c) $126 = (15)(8) + 6$; (d) $256 = (27)(9) + 13$; (e) $369 = (21)(17) + 12$; (f) $1274 = (97)(13) + 13$; (g) $8 = (12)(0) + 8$.
2. $7k, 7k + 1, 7k + 2, 7k + 3, 7k + 4, 7k + 5, 7k + 6$.
8. (a) $12k + 1, 12k + 3, 12k + 5, 12k + 7, 12k + 9, 12k + 11$; (b) $12k + 1, 12k + 2, 12k + 4, 12k + 5, 12k + 7, 12k + 8, 12k + 10, 12k + 11$; (c) $12k + 1, 12k + 5, 12k + 7, 12k + 11$.
9. (a) $6k + 1, 6k + 3, 6k + 5$; (b) $6k, 6k + 2, 6k + 4$; (c) $6k, 6k + 3$.

CHAPTER 2   EXERCISE 4

1. If $a \mid b$ then $ak = b$ where k is an integer. Since $a + b = c$, by the law of substitution $a + ak = c$ and hence $a(k + 1) = c$ and $a \mid c$.
6. Three consecutive integers may be represented by $n, n + 1$, and $n + 2$. If $3 \mid n$, the theorem is true. If $3 \nmid n$, then n is of the form $3k + 1$ or of the form $3k + 2$. If n is of the form $3k + 1$, then $n + 2 = (3k + 1) + 2 = 3k + 3$, which is divisible by 3. If n is of the form $3k + 2$, then $n + 1 = (3k + 2) + 1 = 3k + 3$ which is divisible by 3. Hence the product of three consecutive numbers is divisible by 3.
7. All integers may be written in one of the forms $6k, 6k + 1, 6k + 2, 6k + 3, 6k + 4$, or $6k + 5$. If $n = 6k$, the product is divisble by 6 since n is divisible by 6. If n is of the form $6k + 1$, then $n + 1 = (6k + 1) + 1 = 6k + 2$ which is divisible by 2, and $2n + 1 =$

$2(6k + 1) + 1 = 12k + 3$, which is divisible by 3. Since one of the factors of the product is divisible by 2 and one by 3, the product is divisible by $(2)(3) = 6$. If n is of the form $6k + 2$, then n is divisible by 2 and $n + 1 = (6k + 2) + 1 = 6k + 3$ which is divisible by 3. Again since one of the factors of the product is divisible by 2 and one is divisible by 3, the product is divisible by $(2)(3) = 6$. Similarly, it can be shown that if n is of the form $6k + 3$, $6k + 4$, or $6k + 5$, the product is divisible by 6. The reader should complete the proof.

CHAPTER 2   EXERCISE 4

1. (a); (b); (d); (e); (g).
2. (b); (d); (g); (h).
3. (c); (d); (e).
4. (a); (d).
5. (g).
6. (a); (c); (f).
7. (a); (c); (d); (f).
10. (a); (c); (f).

CHAPTER 3   EXERCISE 1

2. 2.
3. No; the number named would be even and hence divisible by 2.
4. No; the number named would be divisible by 5.
5. 1, 3, 7, 9.
6. (a) 211; (b) 2311; (c) 30031; (d) 223092871.
7. (a) 47, 19; (b) 5, 73; (c) 3, 11, 101; (d) 2, 5, 73.
14. (c); (d); (e).

CHAPTER 3   EXERCISE 2

1. (a) 2, 3, 13; (b) 5, 73; (c) 2, 101; (d) 2, 3; (e) 2, 3, 5; (f) 2, 211.
2. (a) 3×31; (b) $2^3 \times 3^2$; (c) 2×59; (d) $2^2 \times 47$; (e) 7×101; (f) 5×691.
4. (a) 173; (b) 31; (c) 101; (d) 7.
5. 1 and itself.

CHAPTER 3   EXERCISE 3

1. (a) 8; (b) 11.
2. (a) 2; (b) 1.
3. (a) 17; (b) 4.
5. $x = 4, y = -17$; there are many answers.
6. $x = -9, y = 25$; there are many answers.
7. $x = 6, y = -17$; there are many answers.
8. $x = -3, y = 4$; there are many answers.
9. $x = 42, y = -55$; there are many answers.
11. (b); (c); (f); (h).
12. 1.
13. n.

CHAPTER 3   EXERCISE 4

1. 137, 139; 149, 151; 179, 181; 191, 193; 197, 199; 227, 229.
2. (a) 3 + 17, 7 + 13; (b) 7 + 23, 13 + 17; (c) 3 + 29, 13 + 19; (d) 4 + 47.
7. 13 − 3; 17 − 7; 23 − 13.
8. (a) 3; (b) 13; (c) 23; (d) 97.
9. 29, 31, 37, 41, 47.

CHAPTER 4   EXERCISE 1

1. (a) 5; (b) 0; (c) 3; (d) 5; (e) 4; (f) 5.
2. (a) 5; (b) 4; (c) 1; (d) 6.
3. (a) 1; (b) 5; (c) 3; (d) 1; (e) 5; (f) 6.
4. (a) 2; (b) 5; (c) 2; (d) 6.
6. Modulo-five system.
9. (c) The additive inverse of 0 is 0; of 1, 4; of 2, 3; of 3, 2; of 4, 1;
   (d) The multiplicative inverse of 1 is 1; of 2, 3; of 4, 4; of 3, 2;
   (e) 4; (f) 4.
11. (a) 6; (b) It has none; (c) No; 0, 2, and 4; (d) No solution.

CHAPTER 4   EXERCISE 2

2. (a) 4; (b) 4; (c) 5; (d) 3; (e) 6; (f) 4.
3. (a) 6; (b) 4; (c) 0; (d) 4; (e) 0; (f) 8.
5. Sixteen.

CHAPTER 4   EXERCISE 3

1. (b); (c); (d).
2. (a) 6; (b) 5, 8; (c) 4, 7; (d) 5; (e) 5, 7.
3. (a) None; (b) 5; (c) None; (d) 6; (e) None; (f) None.
5. (a) 3; (b) 2; (c) 0, 2, 4, 6; (d) 1, 3, 5, 7.
7. (a) 8; (b) 9; (c) 2, 6, 10.
8. (a) 5; (b) 1, 10; (c) 8.

CHAPTER 4   EXERCISE 4

4. (b); (e).
5. (a) 1; (b) 5; (c) 6; (d) 6.
6. $\{\ldots, -6, -3, 0, 3, 6, \ldots\}$.
7. (a) 6, 17, 28, 39, 50; (b) −5, 12; (c) −33, −15, 3, 31, 39.
8. (a) $x \equiv 7 \pmod{11}$; (b) $x \equiv 3 \pmod 9$; (c) $x \equiv 5 \pmod{23}$; (d) $76x \equiv 25 \pmod{101}$.
9. 0.
12. (a) $4x \equiv 5 \pmod 9$; (b) $9x \equiv 13 \pmod 4$ or $x \equiv 1 \pmod 4$; (c) $4x \equiv 7 \pmod 3$ or $x \equiv 1$ (mod 3); (d) $9x \equiv 4 \pmod 6$ or $3x \equiv 4 \pmod 6$.

CHAPTER 4   EXERCISE 5

1. (a) 1; (b) 1, 2; (c) No solution; (d) No solution; (e) 0, 2, 4, 6.
2. (a) No solution; (b) No solution; (c) 2, 3; (d) 1; (e) 2.
3. (a) No solution; (b) 4; (c) 2, 5; (d) 0, 3, 6; (e) No solution.
4. (a) No solution; (b) No solution; (c) 7; (d) 9; (e) 1, 3, 5.

CHAPTER 4   EXERCISE 6

1. (a); (b); (d); (e).
2. (a) 1; (b) 2; (c) None; (d) 9; (e) 1.
3. (a) $x \equiv 2 \pmod{24}$, $x \equiv 10 \pmod{24}$, $x \equiv 18 \pmod{24}$; (h) $x = 10 \pmod{25}$; (c) No solution.
4. (a) $x \equiv$ 4, 9, 14, 19, 24, 29, 34, 39, 44, 49, 54, 59, 64, 69, 74, 79, 84, 89, 94, 99, 104, 109, 114, 119, 124, 129, 134, 139, 144, 149, 154, 159, 164, 169, 174, 179, 184, 189, 194, 199, 204, 209, 214, 219, 224, 229, 234, 239, 244, 249, 254, 259, 264, 269, 274, 279, 284, 289, 294, 299, 304, 309, 314, 319, 324, 329, 334, 339, 344, 349, 354, 359, (mod 360); (b) $x \equiv 51 \pmod{360}$; (c) No solution.
5. (a) No solution; (b) $x \equiv$ 0, 4, 8, 12, 16, 20, 24, 28, 32, 36, 40, 44, 48, 52, 56, 60, 64, 68, 72, 76, 80, 84, 88, 92 (mod 96); (c) $x \equiv$ 4, 11, 18, 25, 32, 39, 46, 53, 60, 67, 74, 81, 88, 95, 102, 109, 116, 123, 130, 137, 144, 151, 158, 165, 172 (mod 175).

CHAPTER 5   EXERCISE 1

1. (a); (c); (d); (e); (f).
3. 3.
4. $\sqrt{112.25} \doteq 10.6$.
5. $5\sqrt{5} \doteq 11.2$.
6. $\sqrt{68} \doteq 8.25$.
7. $\sqrt{468} \doteq 21.6$.
8. $AC = \sqrt{20} \doteq 4.5$; $AD = \sqrt{24} \doteq 4.9$.
9. 17.
10. $7\frac{2}{3}$.
14. 51 ft.
15. $\sqrt{52} \doteq 7.2$ ft.
16. $60\sqrt{2} \doteq 84.8$.

CHAPTER 5   EXERCISE 2

1. (e); (f); (h); (i); (j).
2. (a) 4, 3, 5; (b) 12, 5, 13; (c) 40, 9, 41; (d) 84, 13, 85; (e) 176, 185, 57; (f) 800, 369, 881; (g) 400, 561, 689; (h) 252, 275, 373.
4. $x = 220$, $y = 21$, $z = 221$; $x = 20$, $y = 21$, $z = 29$.
5. In this case $b = 1$ and hence $a = 2, 4, 6, 8, \ldots$. If $b = 1$ and $a = 2$, then $x = 4$, $y = 3$, and $z = 5$.
6. (a) $y = 195, z = 197$; $y = 45, z = 53$; (b) $y = 15, z = 17$; (c) $y = 35, z = 37$; $y = 5, z = 13$; (d) $y = 63, z = 65$; (e) $y = 99, z = 101$; $y = 21, z = 29$; (f) $y = 255, z = 257$; (g) $y = 323, z = .325$; $y = 77, z = 85$; (h) $y = 399, z = 401$; $y = 9, z = 41$.

7. (a) $x = 24, z = 25$; (b) $x = 312, z = 313$; (c) $x = 612, z = 613$; $x = 12, z = 37$; (d) $x = 364$, $z = 365$; $x = 36$, $z = 45$; (e) $x = 40$, $z = 41$; (f) $x = 1012$, $z = 1013$; $x = 28$, $z = 53$; $x = 108, z = 117$.

CHAPTER 5   EXERCISE 3

1. $n^2 + 1$.
2. $\sqrt{p + q}$.
4. (a) 10; (b) $\sqrt{52} = 2\sqrt{13}$.
5. $\sqrt{168.75}$.
6. It is not a right triangle.
8. (a) $6\sqrt{2}$; (b) $9\sqrt{2}$; (c) 2; (d) $13\sqrt{2}$; (e) $2\sqrt{3}$; (f) $17\sqrt{2}$.

CHAPTER 6   EXERCISE 1

1. (a) 465, 40484; (b) 1042, 71632; (c) 280, 9399; (d) 143, 4872.
2. (a); (c).
3. (c).
4. (a) Reflexive, symmetric, transitive; (b) Symmetric, transitive; (c) Reflexive, symmetric, transitive.
5. (a) Transitive; (b) Transitive; (c) Reflexive, symmetric, transitive.
6. (a) None; (b) Transitive; (c) Reflexive, symmetric; (d) None.

CHAPTER 6   EXERCISE 2

7. (a) Even + (odd + odd) = even + even
$$= \text{even}.$$
(Even + odd) + odd = odd + odd
$$= \text{even}.$$
∴ Even + (odd + odd) = (even + odd) + odd.
(b) Even × (odd + odd) = even × even
$$= \text{even}.$$
(Even × odd) + (even × odd) = even + even
$$= \text{even}.$$
∴ Even × (odd + odd) = (even × odd) + (even × odd).

CHAPTER 6   EXERCISE 3

1. (a) $\frac{5}{3}$; (b) $\frac{4}{3}$; (c) $-\frac{6}{7}$; (d) $-\frac{7}{6}$; (e) $\frac{15}{-7}$; (f) $\frac{12}{19}$.
2. (a) $-\frac{1}{2}$; (b) $-\frac{3}{4}$; (c) $\frac{1}{4}$; (d) $\frac{2}{3}$; (e) $\frac{1}{2}$; (f) $\frac{7}{-9}$.
3. (a) 1, 2, 3, 4, 6, 7, 8, 9, 11; (b) 1, 2, 3, 4, 6, 7, 8, 11; (c) 2, 3, 6, 7, 8, 9, 11; (d) 1, 2, 3, 7, 8, 11; (e) All; (f) 2, 3, 4, 5, 7, 8, 9, 11; (g) 1, 2, 3, 4, 5, 6, 7, 8, 11; (h) 1, 2, 3, 4, 5, 6, 7, 8, 11.

CHAPTER 6   EXERCISE 4

1. No inverses, no identity.
2. No inverses.

3. (e); (f); (g).
4. (a) Yes; (b) The inverse of 1 is 1; the inverse of 2 is 3; the inverse of 3 is 2; the inverse of 4 is 4.

CHAPTER 6   EXERCISE 5

1. (a) V; (b) V; (c) D'; (d) R_3; (e) H; (f) R_0.

3.

*	R_0	R_1	R_2	V_1	V_2	V_3
R_0	R_0	R_1	R_2	V_1	V_2	V_3
R_1	R_1	R_2	R_0	V_2	V_3	V_1
R_2	R_2	R_0	R_1	V_3	V_1	V_2
V_1	V_1	V_3	V_2	R_0	R_2	R_1
V_2	V_2	V_1	V_3	R_1	R_0	R_2
V_3	V_3	V_2	V_1	R_2	R_1	R_0

$R_0, R_1, R_2, V_1, V_2,$ and V_3 are defined as shown below:

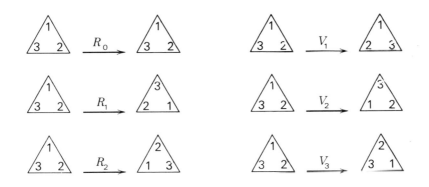

This system is a group, but not a commutative group.

CHAPTER 6   EXERCISE 6

2.

$a * x = a$	Hypotheses
$a * e = a$	Identity Postulate
$a * x = a * e$	Transitive Property of the Equality Relation
$a^{-1} * (a * x) = a^{-1} * (a * e)$	Closure and Inverse Postulates
$(a^{-1} * a) * x = (a^{-1} * a) * e$	Associative Property
$e * x = e * e$	Inverse Postulate
$x = e$	Identity Postulate

4. We want to prove that $(a * b) * (b^{-1} * a^{-1}) = e$.

$(a * b) * (b^{-1} * a^{-1}) = [a * (b * b^{-1})] * a^{-1}$	Associative Property
$= (a * e) * a^{-1}$	Inverse Postulate
$= a * a^{-1}$	Identity Postulate
$= e$	Inverse Postulate

CHAPTER 6   EXERCISE 7

1.

*	e	p	q	r	s	t
e	e	p	q	r	s	t
p	p	q	e	s	t	r
q	q	e	p	t	r	s
r	r	t	s	e	q	p
s	s	r	t	p	e	q
t	t	s	r	q	p	e

2. (a) s; (b) e; (c) t; (d) q; (e) r; (f) q.

3. e.

4. The inverse of e is e; the inverse of p is q; the inverse of q is p; the inverse of r is r; the inverse of s is s; and the inverse of t is t.

5. (a) 2; (b) 120; (c) 24.

6.

*	e	p
e	e	p
p	p	e

9. $\begin{pmatrix} A & B & C & D \\ B & D & C & A \end{pmatrix}$

CHAPTER 7   EXERCISE 1

1. {Heads, tails}.

2. (a) $\frac{1}{2}$; (b) $\frac{1}{2}$; (c) $\frac{1}{2}$; (d) $\frac{1}{2}$.

3. (a) $\frac{1}{6}$; (b) $\frac{1}{6}$; (c) $\frac{1}{2}$; (d) $\frac{2}{3}$; (e) $\frac{1}{2}$.

4. $\frac{1}{6}$.

5. 52.

6. $\frac{1}{52}$.

7. $\frac{1}{13}$.

8. (a) $\frac{1}{4}$; (b) $\frac{3}{13}$; (c) $\frac{1}{13}$.

9. No; $\frac{3}{5} + \frac{1}{4} \neq 1$.

10. $\frac{3}{4}$.

11. (a) $\frac{1}{10}$; (b) $\frac{13}{50}$; (c) $\frac{3}{5}$.

12. (a) $\frac{3}{8}$; (b) $\frac{1}{4}$; (c) $\frac{3}{16}$; (d) $\frac{1}{8}$; (e) $\frac{1}{16}$.

13. (a); (b); (d).

CHAPTER 7   EXERCISE 2

1. (a) $\frac{1}{36}$; (b) $\frac{5}{36}$; (c) $\frac{1}{12}$; (d) $\frac{5}{36}$; (e) $\frac{1}{2}$; (f) $\frac{13}{18}$.

2. (a) HHHH, HHHT, HHTH, HHTT, HTHH, HTHT, HTTH, HTTT, THHH, THHT, TTHH, TTHT, TTTH, TTTT, THTT, THTH; (b) $\frac{1}{16}$; (c) $\frac{1}{4}$; (d) $\frac{1}{16}$; (e) $\frac{3}{8}$.

3. (a) H1, H2, H3, H4, H5, H6, T1, T2, T3, T4, T5, T6; (b) $\frac{1}{12}$; (c) $\frac{1}{2}$; (d) $\frac{1}{6}$.

4. 5 heads; 5 tails; 4 heads, 1 tail; 3 heads, 2 tails; 2 heads, 3 tails; 1 head, 4 tails.

5. $\frac{1}{52}$.

6. (a) Yes; (b) $\frac{1}{10}$; (c) $\frac{3}{20}$; (d) 1.

7. (a) 16; (b) $\frac{1}{16}$; (c) $\frac{1}{4}$; (d) 64, $\frac{3}{64}$.
8. (a) $\frac{1}{2}$; (b) $\frac{1}{2}$; (c) $\frac{1}{5}$.
9. (a) $\frac{7}{64}$; (b) $\frac{1}{64}$; (c) $\frac{3}{32}$; (d) $\frac{3}{32}$.
10. $\frac{1}{3}$.

CHAPTER 7   EXERCISE 3

1. (a); (c); (e).
2. (a) $\frac{1}{12}$; (b) $\frac{5}{18}$; (c) $\frac{5}{18}$; (d) $\frac{1}{6}$; (e) $\frac{1}{12}$.
3. (a) $\frac{5}{11}$; (b) $\frac{2}{11}$; (c) $\frac{6}{11}$; (d) $\frac{9}{11}$; (e) 0.
4. $\frac{39}{1000}$.
5. (a) $\frac{12}{29}$; (b) $\frac{4}{29}$; (c) $\frac{16}{29}$; (d) $\frac{13}{29}$.
6. (a) $\frac{1}{20}$; (b) $\frac{1}{2}$; (c) $\frac{1}{2}$.
7. $\frac{3}{4}$.
8. $\frac{1}{36}$.
9. $\frac{1}{3}$.
10. $\frac{3}{5}$.
11. 1.

CHAPTER 7   EXERCISE 4

1. $P(X \cap Y \cap Z \cap W) = P(X) \cdot P(Y) \cdot P(Z) \cdot P(W)$; (b) $(\frac{1}{2})(\frac{1}{2})(\frac{1}{2})(\frac{1}{2}) = \frac{1}{16}$.
2. (a); (b); (e).
3. $\frac{1}{169}$.
4. (a) $\frac{1}{8}$; (b) $\frac{1}{24}$; (c) $\frac{1}{24}$; (d) $\frac{1}{24}$.
5. (a) $\frac{2}{9}$; (b) $\frac{4}{9}$; (c) $\frac{1}{5}$; (d) 1.

CHAPTER 7   EXERCISE 5

1. (a) Mean 5; median 4; mode 3; (b) Mean 23; median 22.5; mode 21; (c) Mean 3; median 3; mode 3.
2. 5.5.
3. 1464.
4. 14.37.
5. 88.
6. 24,475.
7. (a) $7500; (b) $6000.
8. (a) 124; (b) 128; (c) 128.
9. (a) $c\overline{X}$; (b) $\overline{X} + k$.

CHAPTER 7   EXERCISE 6

1. Mode.
2. (a) $7123.08; (b) $5000; (c) $5000.
4. Mode.
5. (a) Mean 73.55, median 25, mode 25.

CHAPTER 7   EXERCISE 7

1. (a) 4.3; (b) 2.
2. Mean 76, median 80, mode 80, standard deviation 16.0.
3. Mean 3.01, standard deviation 0.04.
5. 2.
6. (a) Fred, Arthur; (b) Arthur; (c) Lela, Gayle, Walter; (c) Jan.

CHAPTER 8   EXERCISE 1

1. (a) 1. Each pair of wires in S has at least one bead in common.
    2. Each pair of wires in S has not more than one bead in common.
    3. Every bead in S is on at least two wires.
    4. Every bead in S is on not more than two wires.
    5. The total number of wires is four.
2. (a) 1. Each pair of committees in S has at least one member in common.
    2. Each pair of committees in S has not more than one member in common.
    3. Every member in S is on at least two committees.
    4. Every member in S is on not more than two committees.
    5. The total number of committees is four.
3. (a) 1. Each pair of rows in S has at least one tree in common.
    2. Each pair of rows in S has not more than one tree in common.
    3. Every tree in S is in at least two rows.
    4. Every tree in S is in not more than two rows.
    5. The total number of rows is four.
4. (a) 1. Each pair of galaxies in S has at least one star in common.
    2. Each pair of galaxies in S has not more than one star in common.
    3. Every star in S is in at least two galaxies.
    4. Every star in S is in not more than two galaxies.
    5. The total number of galaxies is four.

CHAPTER 8   EXERCISE 2

1. Club A: May, James, Sue; Club B: May, Tom, Bob; Club C: James Tom, Alice; Club D: Sue, Bob, Alice; Parallel pairs: May-Alice; James-Bob; Sue-Tom.
3. A is parallel to F; B is parallel to E; C is parallel to D.
4. 1. Each pair of points in S has at least one line in common.
    2. Each pair of points in S has at most one line in common.
    3. Every line in S goes through at least two points.
    4. Every line in S goes through not more than two points.
    5. The total number of points in S is four.
6. Theorem 8.1: There will be exactly six lines in S.
    Theorem 8.2: There are exactly three lines going through each point in S.
    Theorem 8.3: Each line in S has one and only one line parallel to it.
    Corollary 8.1: On a point in S, not on a given line in S, there is one and only one line parallel to a given line.
    Definition 8.1: Two lines which have no points in common are called parallel lines.
7. Postulates 1 and 2.
8. Six lines and four points.
9. Postulates 1 and 2.

CHAPTER 8   EXERCISE 3

1. To show the independence of Postulate 2, let us interpret S to consist of seven points, A, B, C, D, E, F, and G, distributed among four lines, I, II, III, and IV as follows:

I	A	B	C	G
II	A	D	E	G
III	B	F	D	
IV	C	F	E	

In this interpretation Postulate 2 fails because lines I and II have more than one point in common. Postulate 1 is satisfied because each pair of lines has at least one point in common. Postulate 3 is satisfied because every point is on at least two lines. Postulate 4 is satisfied because every point is on at most two lines. Postulate 5 is satisfied because there are four lines.

2. To show the independence of Postulate 4 let us interpret S to consist of four points, A, B, C, and D, distributed among four lines, I, II, III, and IV, as follows:

I	A	B	D
II	A	C	
III	B	C	
IV	C	D	

In this interpretation Postulate 4 fails because point C is on more than two lines. Postulate 1 is satisfied because each pair of lines has at least one point in common. Postulate 2 is satisfied because each pair of lines has at most one point in common. Postulate 3 is satisfied because every point is on at least two lines. Postulate 5 is satisfied because there are four lines.

3. To show the independence of Postulate 3 let us interpret S to consist of seven points, A, B, C, D, E, F, and G, distributed among four lines, I, II, III, and IV, as follows:

I	A	B	C	
II	A	D	E	
III	B	E	F	G
IV	C	D	G	

In this interpretation Postulate 3 fails because each point is not on two lines. Postulate 1 is satisfied because each pair of lines has at least one point in common. Postulate 2 is satisfied because each pair of lines has not more than one point in common. Postulate 4 is satisfied because every point is on not more than two lines. Postulate 5 is satisfied because there are four lines in S.

4. To show the independence of Postulate 5 let us interpret S to consist of three lines, I, II, and III, and three points, A, B, and C, distributed as follows:

I	A	B
II	A	C
III	B	C

In this interpretation Postulate 5 fails because there are only three lines in S. Postulate 1 is satisfied because each pair of lines has at least one point in common. Postulate 2 is satisfied because each pair of lines has at most one point in common. Postulate 3 is satisfied because every point in S is on at least two lines. Postulate 4 is satisfied because every point in S is on at most two lines.

CHAPTER 8   EXERCISE 4

1.

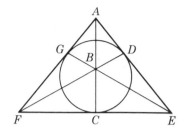

3. $L_1 : P_2, P_3, P_4$; $L_2 : P_1, P_3, P_5$; $L_3 : P_1, P_2, P_6$; $L_4 : P_1, P_4, P_7$; $L_5 : P_2, P_5, P_7$; $L_6 : P_3, P_6,$ P_7 ; $P_7 : P_4, P_5, P_6$.

CHAPTER 8   EXERCISE 5

1. Theorem 8.6 : Two lines intersect in one point.
   Theorem 8.7 : Two points determine one line.
   Theorem 8.8 : If L is any line, there is at least one point that does not lie on L.
2. Postulate 1 : If P_1 and P_2 are any two politicians in S, there is at least one committee containing both P_1 and P_2.
   Postulate 2 : If P_1 and P_2 are any two politicians in S, there is at most one committee containing both P_1 and P_2.
   Postulate 3 : If L_1 and L_2 are any two committees in S, there is at least one politician who is on both L_1 and L_2.
   Postulate 4 : There are exactly three politicians on each committee.
   Postulate 5 : If L is any committee in S, there is at most one politician who is not on L.
   Postulate 6 : There exists at least one committee.
3. Theorem 8.4 : There exists at least one politician.
   Theorem 8.5 : If L_1 and L_2 are any two committees, there is at most one politician who is on both L_1 and L_2.
   Theorem 8.6 : Two politicians determine exactly one committee.
   Theorem 8.7 : Two committees have exactly one politician in common.
   Theorem 8.8 : If P is any politician, there is at least one committee that does not have P as a member.
   Theorem 8.9 : Every politician is on at least three committees.
4. This theorem follows directly from Postulates 1 and 2.

CHAPTER 9   EXERCISE 1

1. (b); (c).
2. (a) 7; (b) 6; (c) $\frac{1}{8}$; (d) 2.
3. (a) 3 × 2; (b) 4 × 4; (c) 1 × 4; (d) 2 × 4.
4. (a) 10; (b) 0; (c) 14; (d) B; (e) 35; (f) 37; (g) 5; (h) 14.

5. $\begin{pmatrix} 2 & 4 & 2 \\ 5 & 1 & 1 \\ 6 & 6 & 4 \end{pmatrix}$

6. (a) 0.3; (b) 0.6; (c) 0.3.

7. (a) 15; (b) 5; (c) 2; (d) 6; (e) 4.

CHAPTER 9 EXERCISE 2

1. (a) $\begin{pmatrix} -2 & -4 \\ 11 & 18 \end{pmatrix}$ (b) $\begin{pmatrix} 58 & -16 \\ 177 & -167 \end{pmatrix}$

2. (a) $\begin{pmatrix} \frac{4}{3} & \frac{11}{10} \\ \frac{5}{8} & \frac{17}{12} \end{pmatrix}$ (b) $\begin{pmatrix} 0.2 & 8.5 \\ -1.5 & -9.3 \end{pmatrix}$

3. (a) $\begin{pmatrix} \frac{3}{2} & \frac{3}{2} \\ \frac{1}{8} & \frac{9}{7} \end{pmatrix}$ (b) $\begin{pmatrix} 0 & 15 \\ 0 & 24 \end{pmatrix}$

4. (a) $\begin{pmatrix} 0 & 0 \\ 0 & 42 \end{pmatrix}$ (b) $\begin{pmatrix} \frac{5}{4} & 3\sqrt{3} \\ \sqrt{6} & -4\sqrt{8} \end{pmatrix}$

5. (a) $x = 2, z = 12$; (b) $x = 3, y = -8, z = -7, w = -21$; (c) $x = -5, y = 4, z = -12,$ $w = 84$.

6. (a) $\begin{pmatrix} -7 & -6 \\ -3 & -4 \end{pmatrix}$ (b) $\begin{pmatrix} 1.8 & -1.7 \\ -3.9 & 4.2 \end{pmatrix}$

   (c) $\begin{pmatrix} \frac{1}{2} & \frac{1}{4} \\ -8 & 7 \end{pmatrix}$ (d) $\begin{pmatrix} 0 & 0 \\ 7.6 & -8.7 \end{pmatrix}$

7. (a) $\begin{pmatrix} 12 & 9 \\ 10 & 8 \end{pmatrix}$ (b) $\begin{pmatrix} 12 & 9 \\ 10 & 8 \end{pmatrix}$

   (c) $\begin{pmatrix} -4 & -1 \\ -6 & 5 \end{pmatrix}$ (d) $\begin{pmatrix} 14 & 7 \\ -12 & -5 \end{pmatrix}$

8. (a) $x = 4, y - 5, z = -11, w = 3$; (b) $x = -6, y = 22, z = 3, w = 5$; (c) $x = \frac{1}{4}, y = \frac{1}{4},$ $z = \frac{7}{6}, w = \frac{3}{5}$; (d) $x = 1.2, y = 1.3, z = 4.8, w = 1.4$; (e) $x = \frac{9}{8}, y = 1, z = \frac{15}{8}, w = \frac{3}{16}$; (f) $x = 2.22, y = 0.60, z = 13.13, w = -11.17$.

9. (a) $\begin{pmatrix} 850 & 550 \\ 1051 & 750 \end{pmatrix}$

   (b) 1900; (c) 1400; (d) 1300; (e) 1800.

10. $\begin{pmatrix} 850 & 1075 \\ 730 & 820 \end{pmatrix}$

11. $\begin{pmatrix} 1850 & 1175 \\ 3925 & 2100 \end{pmatrix}$

CHAPTER 9 EXERCISE 3

1. (a) $\begin{pmatrix} 21 & 37 \\ 75 & 133 \end{pmatrix}$ (b) $\begin{pmatrix} 0 & 31 \\ 8 & 98 \end{pmatrix}$

2. (a) $\begin{pmatrix} 4 & 7 \\ 3 & -1 \end{pmatrix}$ (b) $\begin{pmatrix} -8 & 1 \\ -7 & 4 \end{pmatrix}$

3. (a) $\begin{pmatrix} 43 & -35 \\ 40 & 28 \end{pmatrix}$ (b) $\begin{pmatrix} -66 & 233 \\ -8 & 38 \end{pmatrix}$

4. (a) $\begin{pmatrix} -3 & -4 \\ 6 & -10 \end{pmatrix}$ (b) $\begin{pmatrix} 0 & 0 \\ 0 & 0 \end{pmatrix}$

5. (a) $\begin{pmatrix} -3 & -7 \\ -60 & 22 \end{pmatrix}$ (b) $\begin{pmatrix} -80 & -290 \\ -140 & -510 \end{pmatrix}$

6. (a) $\begin{pmatrix} 23 & 27 \\ 16 & 34 \end{pmatrix}$ (b) $\begin{pmatrix} -12 & -52 \\ -6 & -26 \end{pmatrix}$

7. (a) $\begin{pmatrix} 17 & 32 \\ 16 & 33 \end{pmatrix}$ (b) $\begin{pmatrix} 115 & 228 \\ 114 & 229 \end{pmatrix}$

CHAPTER 9   EXERCISE 4

1. (a) $\begin{pmatrix} 1 & -\frac{1}{2} \\ -\frac{1}{2} & \frac{1}{2} \end{pmatrix}$ (b) $\begin{pmatrix} \frac{2}{5} & -\frac{1}{5} \\ -\frac{1}{5} & \frac{3}{5} \end{pmatrix}$

2. (a) $\begin{pmatrix} -1 & 2 \\ 1 & -1 \end{pmatrix}$ (b) $\begin{pmatrix} \frac{1}{2} & 0 \\ -\frac{3}{8} & \frac{1}{4} \end{pmatrix}$

3. (a) $\begin{pmatrix} 1 & -\frac{1}{2} \\ -\frac{1}{2} & \frac{1}{2} \end{pmatrix}$ (b) $\begin{pmatrix} \frac{1}{2} & \frac{1}{2} \\ \frac{1}{4} & \frac{3}{4} \end{pmatrix}$

6. (a), (d) have inverses; (b), (c) do not have inverses.
7. (a), (d) have inverses; (b), (c) do not have inverses.

CHAPTER 10   EXERCISE 1

1. (a) $(7 \times 10^1) + (8 \times 10^0)$; (b) $(1 \times 10^2) + (6 \times 10^1) + (3 \times 10^0)$; (c) $(8 \times 10^2) + (2 \times 10^1) + (7 \times 10^0)$; (d) $(7 \times 10^3) + (6 \times 10^2) + (1 \times 10^1) + (9 \times 10^0)$; (e) $(8 \times 10^4) + (2 \times 10^3)$; (f) $(1 \times 10^5) + (3 \times 10^4)$.
3. (a) 4; (b) 8; (c) 33; (d) 171; (e) 351; (f) 1024.
5. (a) 1000; (b) 1101; (c) 100100; (d) 1000101; (e) 1001101; (f) 1100010.
7. (a) 111; (b) 10010; (c) 1110; (d) 10110.
9. (a) 1000; (b) 1; (c) 1001.

CHAPTER 10   EXERCISE 2

1. (a) 100; (b) 511; (c) 519; (d) 1726.
3. Eight.
4. (a) $5 \times \text{eight}^2$; (b) $5 \times \text{eight}$; (c) $5 \times \text{eight}^3$; (d) $5 \times \text{eight}^4$.
5. (a) 1411; (b) 4423; (c) 21270; (d) 22267; (e) 100000; (f) 100001.
7. (a) 531; (b) 2424; (c) 10640; (d) 7730; (e) 15674; (f) 214500.

CHAPTER 10   EXERCISE 3

1. 1, 10, 11, 100, 101, 110, 111, 1000, 1001, 1010, 1011, 1100, 1101, 1110, 1111, 10000, 10001, 10010, 10011, 10100.
2. 1, 2, 3, 4, 5, 6, 7, 10, 11, 12, 13, 14, 15, 16, 17, 100, 101, 102, 103, 104.
3. (a) 11; (b) 52; (c) 77; (d) 444; (e) 353; (f) 115; (g) 207; (h) 636.
4. (a) 67; (b) 555; (c) 364; (d) 712; (e) 253; (f) 413.
5. (a) 1, 010, 011, 100; (b) 111, 100, 110, 010; (c) 101, 111, 011, 001; (d) 11, 110, 010, 100, 100; (e) 111, 111; (f) 1, 000, 000; (g) 101, 110, 010, 000, 000, 000, 110, 011; (h) 100, 100, 111, 110, 110.
6. (a) >; (b) >; (c) >; (d) <; (e) <; (f) <.

CHAPTER 11   EXERCISE 1

1. (a) Real; (e) integer; (g) integer; (k) real; (m) integer; (q) integer; (t) real.
2. (a) A; (c) IOH; (f) BIG; (h) NEGNTY.
3. (a) An integer cannot contain a decimal point; (c) More than six characters in the name; (e) Commas are not valid.
4. (a); (b); (d); (g); (i); (j); (k) (l).
5. 7 through 72.
7. 6.

CHAPTER 11   EXERCISE 2

1. (a) $X = A + B * (C - D **2)$; (b) $X = A + A * C / (B * D)$.
2. (b) $Y = ((A - (B + C)) / 16.0) * A * B **2$.
3. (a) $Y = A * B / 2.0 + C * D / 2.0$.
4. (b) $Z = A **2 / (1.0 + (A / B) / (C / D))$.
5. (a) $a = 3.1416r^2$; (c) $x = \left(\dfrac{a \cdot b}{c \cdot d}\right)^2$.
6. (b) $a = \dfrac{1.0}{17.0 + a/b}$.
7. (a) $x = 0.0$.
8. (b) $x = 0.5$.
9. (a) $x = 10.0$.
10. (b) $x = 0.33333$.

CHAPTER 11   EXERCISE 3

1. 8.
3. 9.
5. 32.
7. 0.
9. 1.
11. 0.
13. 4.
15. 13.

CHAPTER 11   EXERCISE 4

1. Transfer to statement 100.
3. $X = 70.0$.
5. ICOUNT is not less than or equal to 20, therefore the next sequential statement is executed where ICOUNT = ICOUNT − 1.
6. Since IND = 2, the transfer takes the positive branch to statement number 30 where X is set equal to Y **2.
9. The expression left of the logical operator reduces to −0.55 which is less than 25, therefore X is squared, giving the value $X = 231.04$.

11.   If (SUDS − BIG) 10, 10, 20
    10   SOAP = X
    20   ...

13. $X = 65.0$
        IF (IF + 5) 3, 5, 5
     3   $X = X − 2.0$
        GO TO 7
     5   $X = X + 2.0$
     7   SUM = SUM + X

15.   IF (HRS − 40.0) 60, 60, 75
    60   PAY = RATE * 40.0
        GO TO 100
    75   PAY = RATE * 40.0 + (HRS − 40.0) * RATE * 1.5
   100   ...

CHAPTER 11   EXERCISE 5

2.

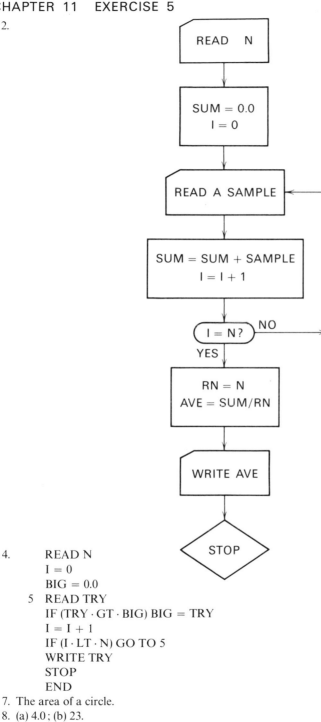

4.   READ N
          I = 0
          BIG = 0.0
       5 READ TRY
          IF (TRY · GT · BIG) BIG = TRY
          I = I + 1
          IF (I · LT · N) GO TO 5
          WRITE TRY
          STOP
          END

7. The area of a circle.

8. (a) 4.0; (b) 23.

CHAPTER 11   EXERCISE 6

1. (a) DIMENSION A(6), B(4), C(10); (b) DIMENSION X(17), I(20); (c) DIMENSION P(10), R(2, 2), S(3, 3).

2.

Y(1)	Y(2)	Y(3)	Z(1, 1)	Z(2, 1)	Z(1, 2)	Z(2, 2)

3. 200.
5. (a) 104; (b) ZEE(1, 1); (c) ZEE(7, 2); (d) ZEE(10, 3); (e) ZEE(5, 6).
7. HIPPY(3, 1), HIPPY(5, 2), HIPPY(12, 2).
9. 239.

CHAPTER 11   EXERCISE 7

```
2.       DIMENSION FP(50)
         SUM = 0.0
         DO 15 L = 1, 16
         SUM = SUM + FP(L)
   15    CONTINUE
4.       DIMENSION SET(40)
         BIG = SET(1)
         DO 60 K = 2, 40
         IF (BIG · LT · SET(K)) BIG = SET(K)
   60    CONTINUE
6.       DIMENSION FIX(30)
         ICOUNT = 0
         DO 3 M = 1, 30
         IF (FIX(M) · LT · 100.0) ICOUNT = ICOUNT + 1
    3    CONTINUE
8.       DIMENSION N(300)
         NPLUS = 0
         NNEG = 0
         NSUMP = 0
         NSUMN = 0
         DO 50 J = 1, 300
         IF (N(J) · GT · 0) GO TO 25
         NNEG = NNEG + 1
         NSUMN = NSUMN + N(J)
         GO TO 50
   25    NPLUS = NPLUS + 1
         NSUMP = NSUMP + N(J)
   50    CONTINUE
         NAVEP = NSUMP / NPLUS
         NAVEN = NSUMN / NNEG
```

Index